Developing National
Power in Space

Developing National Power in Space

A Theoretical Model

BRENT ZIARNICK

McFarland & Company, Inc., Publishers

Jefferson, North Carolina

Library of Congress Cataloguing-in-Publication Data

Ziarnick, Brent David, 1978–
  Developing national power in space : a theoretical model /
Brent Ziarnick.
    p.   cm.
  Includes bibliographical references and index.

  ISBN 978-0-7864-9499-6 (softcover : acid free paper) ∞
  ISBN 978-1-4766-1773-2 (ebook)

  1. Astronautics and state.   2. Astronautics and civilization.
3. Astronautics, Military—United States.   4. Space warfare—
Government policy—United States.   5. Space industrialization.
I. Title.

  TL790.Z53 2015
  629.4—dc23   2014044436

British Library cataloguing data are available

Front cover: Space Shuttle Orbiting Earth
© 2015 3DSculptor/iStock/Thinkstock

Printed in the United States of America

*McFarland & Company, Inc., Publishers
  Box 611, Jefferson, North Carolina 28640
  www.mcfarlandpub.com*

For my wife, Melissa, and my children Ashley, David and Christopher. May this book help you live in a safer and more prosperous future. To the officers and men of the American military space forces past, present, and future. *Ad Astra!*

Table of Contents

Introduction: The Eclipse of American Space Power? 1

1 • The General Theory of Space Power 9

2 • Organizing for Effective Development—Logic 62

3 • Organizing for Effective Development—Grammar 106

4 • The Navalists' War—The Pacific 1941–1945 159

5 • The Spacers' War—Beyond Earth Orbit 2053–2057 201

Chapter Notes 243

Bibliography 249

Index 253

Introduction: The Eclipse of American Space Power?

Man has always known space. From our first ancestor humanity has been able to see the stars and the black void which holds them, simply by looking up. However, only in the last hundred years has mankind been able to harness the power of creation well enough to travel to the heavens. The launch of *Sputnik* on 4 October 1957 forever changed the relationship between space and man. Instead of simply being a source of inspiration, wonder, knowledge, or fear, space became a place where men and his machines can go—a *human environment*. Space travel in the early 21st century is still dangerous, difficult, and expensive, but it is accessible just as man can travel through the air or on and beneath the sea. Man can now *use* space for his own purposes. Man has begun to expand his dominion into space and bring this hostile environment under his control. Man can now build space power.

And at the beginning of the 21st century, the nation that has best harnessed space for its own purposes is the United States. Americans have been the only people to set foot upon another world. Almost every American life is exposed to space services on a daily basis. Drivers reach their destinations through navigation provided by satellite navigation through the Global Positioning System. Weather forecasts are generated using weather satellite data and transmitted to the public through satellite communications. New photos from space telescopes or unmanned probes are constantly posted on the internet and consumed by schoolchildren and the interested public. At any time of the day, basic cable television will undoubtedly be playing a broadcast highlighting space, either a nonfiction history or science program or—more likely—a science fiction adventure. No other society on Earth is as exposed to space as is the United States. No other nation can come close to the amount of wealth and power derived from space as the American nation. But today, many Americans look to the stars unsettled, fearing that they may soon be eclipsed in space by a foreign competitor.

The Paradox

United States space activity, by any measure, is far and away the most advanced and largest program in the world, but there is a growing belief that America has lost its leadership in space to China. Thus the paradox of space power in the early 21st century: The most dominant space power in the world is in crippling fear of being dethroned by a program far smaller. In some metrics, it does appear that Chinese activity is beginning to outpace America's. According to the Space Foundation's *The Space Report 2013:*

> In 2012, China continued to outpace the United States in the number of orbital launches, making 19 orbital launch attempts in 2012, all of which were successful. This makes 2012 the second consecutive year in which China has surpassed the United States as the world's second most active launch operator, due primarily to China's accelerating progress in deploying new scientific and communications satellites, and continued deployment of its Beidou satellite navigation constellation. China's 2012 activity also included its fourth crewed mission, Shenzhou 9, which launched on a Long March 2F in June 2012. Chinese officials have stated that they plan to maintain a launch rate of up to 20 missions a year for the foreseeable future. If China reaches this state, it may pull further ahead of the United States, which has conducted an average of 18 launches per year during the past five years.[1]

Orbital rocket launches are perhaps the most visible and spectacular manifestations of a nation's space activity and it is no surprise that some consider China overtaking the United States in number of annual space launches indicates taking the lead in space. However, we must remember that the world's launch leader for many years has not been the United States, but Russia. Of course, the reasons the Russians launch so many rockets is because they are both a low-cost launch services exporter that deals with many commercial payloads from around the world, and that their national security satellite systems are designed with relatively short service lives, necessitating multiple launches to produce the same level of general service that American systems can accomplish with a single satellite over a decade or longer. The Chinese space program is growing, but simple greater launch rates do not a space leader make.

Even if launch rates don't justify a declaration of a China lead in space, certainly their human spaceflight program merits special consideration. Again, *The Space Report* states the facts:

The fourth Chinese human spaceflight mission, Shenzhou 9, took place in June 2012, achieving several new milestones for China. The primary goal of the Shenzhou 9 mission was to dock with the Tiangong-1 space station, a technology testbed and the first in a series of similar space stations of increasing complexity, designed to eventually lead to a larger, more permanent, modular Chinese space station. While China had conducted automated docking procedures between Tiangong–1 and the unmanned Shenzhou 8 mission in 2011, Shenzhou 9 was the first manually controlled docking operation for China.... So far, China has used each of its four crewed flights to develop its capabilities and to test procedures, in a manner reminiscent of the U.S. Gemini program. This pattern is expected to continue for the foreseeable future.[2]

Alternatively, the report says of the American space program:

> The United States, after the Soviet Union, was the second nation to send a human into space, but it will not have its own human spaceflight capability for the next several years, following the retirement of the Space Shuttle in 2012.[3]

The rapid ascent of the Chinese manned spaceflight program, coupled with the retirement of the American space shuttle, has made a large gap between Chinese and American space capabilities appear to be a grim and undeniable reality. This perceived gap has led some to issue dire pronouncements for American space policy. One former U.S. State Department official claims the "atrophying U.S. space program suggests that America will be forced to cooperate or cede the high frontier of space to China forever."[4] One Naval War College professor calls cooperation essential because "[i]t's one way of preventing a scenario of a galactic Wild West in which China has become the world's leader in space."[5] Even some who don't want the United States to cooperate with China in space and advocate competing against them as adversaries are nonetheless in awe of China's apparent lead. Hotel billionaire and American space entrepreneur Robert Bigelow has stated that the United States cannot contest an inevitable Chinese takeover of the Moon and the only way to defeat China in space is to concede the Moon but beat them to Mars![6]

Chinese-American space cooperation may or may not be a worthwhile goal, but approaching cooperation in supplication from a perceived position of weakness, as a fear of Chinese space ascendancy would entail, would probably be detrimental to American interests. But is China really eclipsing the United States in space? Author Erik Seedhouse argues:

> Thanks to its high-profile manned space missions, much of the world perceives China as catching up with the space capabilities of the U.S. In reality, nothing could be further from the truth but, as China continues to accelerate its manned space program, the two nations may eventually approach a critical juncture that will decide whether the U.S. will be considered as the leader in human spaceflight. However, it is highly unlikely the U.S. will abrogate its leadership role in human space-

flight, since this would have strategic consequences beyond the space realm. Equally, the Chinese, bolstered by the media coverage of their successful manned missions, will be determined to maintain their sustained effort and to see their goal of leadership in space through to a successful conclusion.[7]

So which is it? Has China overtaken the United States as the world's presumed space leader? Must the United States be forced to cooperate with China or risk being swept from the stars altogether? Or is the United States still the undisputed master of space power?

This paradox exists because two different measures of comparative space activity tell two very different stories about China and the United States. The first measure, the absolute value (in dollars) of the size of each country's cumulative space program, clearly shows an American space program an order of magnitude larger than China's. However, the annual rate of growth of each program—the second measure—describes a relatively stagnant American space program compared with a Chinese program expanding at an alarming rate. Futron Corporation's annual Space Competitiveness Index offers a great tool with which to evaluate both the absolute size and rate of growth of America's and China's space programs. Futron describes the index:

> Futron's Space Competitiveness Index is a globally-focused analytic framework that defines, measures, and ranks national competitiveness in the development, implementation, and execution of space activity. By analyzing space-related government, human capital, and economic drivers, the SCI framework assesses the ability of a country to undertake space activity, and evaluates its performance relative to peer nations, as well as the global space arena.[8]

Futron's proprietary model attempts to account for both absolute values and rates of change for each nation's space program, but it does allow for direct comparisons. Comparing both countries through Futron's index, the United States received a score of 91.36 and China received a score of 25.65 on a scale of 0–100. According to the SCI, the United States is clearly dominant. However, Futron explains that this isn't the entire story:

> In the 2009, 2010, 2011, and 2012 SCI results, the United States saw a steady erosion of its position in relation to the other nine countries surveyed. The United States experienced a 4 percent decline in its overall score between 2008, when SCI benchmarking began, and 2012.... China has shown the most impressive gains of any nation, with a 41 percent increase relative to its 2008 starting score.[9]

The United States is dominant, but the SCI also shows that the American program is in a measure of decline, while China program is the fastest-improving space program in the world. Futron assessed China's program favorably for many reasons:

China placed 4th for the second year in a row in the Space Competitiveness Index, solidly ahead of Japan but below Russia. China enjoyed the most pronounced relative competitiveness gains of any nation in the 2012 SCI. Only four nations advanced their relative positions over the previous year. China led these countries, with an average competitiveness increase of 2.52 basis points. In addition, China improved its score relative to every other nation in the study. This was primarily driven by the continued success of its launch industry, advancements in its satellite navigation and manned space programs, and new policy pronouncements unveiling its space activity plans over the next five years.[10]

Alternatively, the United States has the most advanced program in existence, but the rest of the world is catching up.

As in previous editions of Futron's Space Competitiveness Index (SCI), the United States remained the highest-ranked country in the 2012 SCI, with a total score of 91.36. However, the gap between the United States and other nations continues to shrink as other nations enhanced their capabilities relative to the United States.... Key factors accounting for changes in the U.S. score included:
- Decline in the ranking of the U.S. in number of annual launches performed, from second place to third, behind China; and
- General expansion of the space activities of most other nations compared to the United States

The countries that made the greatest advancements against U.S. positioning included China, which gained three basis points.... While the U.S. position as the leading spacefaring nation has gradually eroded, the gap between it and other nations remains large: its overall lead over second-place Europe is more than 40 basis points, and its lead over third-place Russia is more than 50 basis points.[11]

Thus, the American space program is strong but it is losing ground to other countries. The American program's rate of change is relatively stagnant, and may even be contracting. The answer to the paradox, then, is that the American program is large and dominant but the rate of change indicates that it is at risk of not being dominant for much longer, causing fear that the "growth giant" of China will become the leader in space in short order.

A Theory of Space Power Development

Ensuring the dominance of the American space program for years to come boils down to just one thing—causing the rate of change in the value of the U.S. space program to improve until it matches or exceeds that of its world space competitors. Simply achieving even a marginally close approximation to the change rate of competitors such as China will be enough to insure U.S. space mastery for years to come by virtue of America's current commanding lead. But growth alone is simply not enough; any space program must increase its technological capabilities in order to develop. As economist Joseph Schumpeter said, "Add successively as many mail coaches as you please, you will never get a railway thereby."[12] Likewise, launch as many microsatellites as you please, you will never get a manned interstellar starship thereby. In order to add true value, in space power as in economics, growth by itself pales in relation to positive development, and positive development is achieved through innovation.

This book is about developing space power. It will present a theoretical model describing how space power is developed and describe what strategies can be implemented to help foster the development of a nation's space power. The theory is based on classical military and economic theories, and supporting chapters derive their historical and strategic lessons often from military history. Therefore, the ideas presented in this book may be far different than what average space enthusiasts read about space. Alternatively, military readers may be jarred to find references to interstellar flight and other "futuristic" ideas analyzed using methods normally reserved for conventional military affairs. The author hopes that by opening new vistas to many communities, a better synthesis of these communities may arise to champion the development of space power.

Lastly, since this book is an attempt to write a serious military-type strategic theory for a nation's space program, it must adhere to the needs of military theory. Navy Admiral J.C. Wylie stressed what theory should do:

> A theory in any such field as that of strategy is not itself something real and tangible; it is not something that actually has concrete existence. A theory is simply an idea designed to account for actuality or what the theorist thinks will come to pass as actuality. It is orderly rationalization of real or presumed patterns of events. A basic measure of validity of any theory is how closely the postulates of the theory coincide with reality in any actual situation. If any military theory has any proven validity, it is because some practicing military man has actually given it that validity in a real situation. The theory serves a useful purpose to the extent that it can collect and organize the experiences and ideas of other men, sort out which of them may have a valid transfer value to a new and different situation, and help the practitioner to enlarge his vision in an orderly, manageable, and useful fashion—and then apply it to the reality with which he is faced.[13]

Admiral Wylie sets the demands on a successful theory. Professor Harold Winton argues that successful theories must accomplish five functions. First, the theory must "define the field of study under investigation." Second, the theory must "categorize—to break the field of study into its constituent parts."[14] Third, and most importantly, the theory must explain its subject. The theory must be able to explain why things happen the way they do. To Winton, "explanation is the soul of theory." Fourth, the theory must connect the field of study to "other related fields in the universe." Lastly, theory must anticipate.[15]

It must reasonably predict future results given a solid understanding of the present facts of an endeavor involving the field of study.

The space power theory described in Chapter 1 will endeavor to meet all of Dr. Winton's requirements. We will define space power and compare this historically derived definition to other definitions offered in the past. We will then break our subject down into its constituent parts: the Logic and Grammar of Space Power and the elements which comprise both parts. Next, the theory will explain how certain activities can generate space power and how space power is developed. Finally, the chapter will connect this space power theory to other fields of study: in this case primarily economic theory, military theory, and political science, and general strategy, in order to show that space power is a human endeavor and answerable to human behavior.

The next four chapters will tackle Winton's final function of theory—anticipation. Chapter 2 will investigate the Grammar of Space Power and develop explanatory tools and concepts that the theory anticipates will increase an agent's space power if employed. Chapter 3 will turn to the Logic of Space Power and anticipate organizational actions that can increase the space power of a nation. Chapter 4 will present a historical example from the U.S. Navy and its approach to sea power development from the 1880s to the end of World War II's Pacific War to showcase the concepts behind this space power model in action. Finally, Chapter 5 will apply the theory to anticipate which technologies and which organizational changes would most benefit an American space force in dealing with a number of potential space power challenges in the middle of this century.

The theory and ideas presented in this book cannot be used and are not intended to be used as a cookbook aiming to explain to space power professionals, policy makers, and other space leaders exactly what to do in any given situation. Rather, this book is intended to fulfill what Winton hoped a mature space power theory could do: to assist in the self-education of space leaders and identify the explanatory relationships that guide the use and development of effective space power as best as it is able.[16] If this book, in some small part, assists in the positive development of the next generation of American space leaders then it will have served its purpose. Perhaps these leaders may even drive American space power development to levels that are the envy of the world ... even the Chinese.

Chapter 1

The General Theory
of Space Power

This chapter outlines a General Theory of Space Power. It is the *general* theory because it intends to describe all space activity for whatever purpose in whatever era. This is in contrast to other space power theories which have tended to focus on *specific* applications of space activity, such as military space operations. By describing space power in its broadest form, the general theory can be applied to any type of space activity, real or imagined. The General Theory demonstrates a complete view of space activity to allow the reader opportunity to see the many elements involved in space power.

The Intent of the General Theory

The General Theory of Space Power intends to accomplish multiple objectives. Space power theory is still a relatively new field and, as yet, most models offered have been incomplete in a number of ways. Some have been driven entirely by recent technology and operations; others have been entirely devoted to military modes of operation. This space power model, an adaptation of Admiral Alfred Thayer Mahan's and economist Joseph Schumpeter's ideas (among others), is offered as an attempt to correct some of these mistakes and expand the analysis to include past, present and future space activities. Specifically, the model intends to accomplish the following objectives:

1. *The General Theory intends to be comprehensive across activity.* The General Theory intends to understand space power and its development in its totality, and consequently must be applicable across all forms of space activity: commercial, civil, political, and military. It

does not focus on any one specific activity and does not unduly prefer military activity. The model is intended to be a key tool for military space planners, but should be equally useful to political and economic space interests. It is able to assess military projects like space-based radars and commercial efforts such as space tourism on an even level and offer guidance on what space endeavors should be promoted by governments to improve the nation's space power. Like Mahan's sea power theory, the General Theory includes all space activities and offers insight into which activities are most valuable for aspiring space powers.

2. *The General Theory intends to be universal across time.* This model is not intended to be limited by technology or time frames past, present or future. Just as Mahan's sea power theory was derived from the Age of Sail but of immediate use in the oil-fueled Pacific War and even today's nuclear age, the General Theory intends to explain actions throughout the duration of the human space effort. Whether used in exploring the Cold War space campaigns, the age of satellites as global utilities, future activities to colonize the Moon, or even interstellar cruisers of science fiction fame, this model is meant to provide a ready and useful framework for analysis.

3. *The General Theory intends to be descriptive.* The General Theory posits the elements of space power and how commercial, political, and military space power interact with each other. Using the General Theory, we can explore space history to find why some space activities succeeded and others failed, and how space powers can rise or fall. Instead of merely mimicking history and assuming success, the General Theory can criticize past actions against a space power ideal. Indeed, the General Theory does not find much to congratulate in history and instead offers that space history is mostly a story of blunders and poor actions as leaders embraced one or more mistakes in making space policy.

4. *The General Theory intends to be prescriptive.* The General Theory offers specific advice on what space powers must do to gain, develop, and keep space power. Thus, the General Theory intends to inform policy makers and provide them with advice to build better strategies and space policies focused on space power growth. It offers an ideal approach, as well as a discussion and analysis framework from which to judge various courses of action. The author believes the space powers that follow the prescriptive advice of the General Theory will

emerge as viable and effective space powers and help lead their people to both security and prosperity through space activity.

5. *The General Theory intends to bridge the gap between military realism and enthusiast futurism.* Sea and air power have always had military officers who were also enthusiasts for developing their environment. Though sea power's beginnings are lost in antiquity, we know that air officers were among the first to call for exotic equipment thought of as science fiction in their day—consider Billy Mitchell's visions of supersonic, high altitude, heavy bombers. Space power does not currently enjoy this continuity. Promotion of lunar bases and manned spaceflight usually come from the National Aeronautics and Space Administration (NASA), not Air Force Space Command. Space enthusiasts dream of living in space, while military space officers are focused on cold short-term realities. In order to grow space power, space officers need to become space enthusiasts, and enthusiasts need to adopt military terminology to better work together to promote their common interests in space development. The General Theory, combining both views into a single continuum, may help begin an essential dialogue.

The General Theory is not a model that pretends to account for every interaction among its constituent parts (i.e., a systems dynamics model) nor does it claim mathematical precision. It is meant only as a qualitative and top-level model that can provide policy makers, strategists, and analysts a visual representation of high-level operations and relationships. Numerous feedback loops occur among all of the model's parts at some level. Also, there is no linear progression between the Grammar and Logic deltas (its two main component parts), but a continuous ebb and flow of multiple technologies and doctrines. Regardless, a linear flow model such as the General Theory sufficiently contains the essence of space power development and is effective as an analysis tool.

The General Theory strives to be of more than simple academic interest. It is meant to be used by policy makers and strategists, enthusiasts and businessmen, and space realists, to help develop and test the validity of their proposed space activities. No model can be perfect or complete without years of debate, study, and peer review, and this General Theory is neither perfect nor complete. As strategist Colin Gray says, "A powerful explanatory tool, which is what good theory should be, need not be capable of explaining everything" in order to be useful.[1]

However, it is hoped that this introduction will serve as a firm base with which to plan and analyze humanity's conquest of the space environment for peace, prosperity, and security.

The Definition of Space Power

> *Air power may be defined as the ability to do something in the air.*
> —Brigadier General William "Billy" Mitchell, *Winged Defense* [2]

Authors have offered many definitions of space power over the years of human space activity, and biases inherent in the writers have made most of the definitions woefully inadequate. Common to many definitions is an unhealthy preoccupation with the military form of space power. Captain Fritz Baier defined space power as "the ability to use spacecraft to create military and political effects" which identified two uses of space power (creating both military and political effects). [3] However, in the next sentence political effects are gone as his alternate definition is stated as "military power that comes from, resides in, or moves through space while performing its mission." [4] Since most "power" theories are primarily interested in military and force applications, it is somewhat understandable that such biases towards armed conflict emerge in definitions, but it is not excusable because it binds both the writer and reader into a severe myopia that can result in truly bizarre statements. For instance, Colonel Philip Meilinger states that air power (an environmental power that has been almost completely subsumed by its military dimension) is "[in] essence ... targeting, targeting is intelligence, and intelligence is analyzing the effects of air operations." [5] Meilinger's primary audience is composed of Air Force officers, so in the proper context his statement makes sense. However, surely the Central Intelligence Agency doesn't consider its intelligence work solely analyzing air operations. Also, are not jet engines an example of air power, too? Meilinger's definition should warn us of the dangers in pinning a definition on too specific an agenda directed at too narrow an audience.

Other definitions of space power have gotten closer to the mark. Lieutenant Colonel David Lupton writes, "space power is the ability of a nation to exploit the space environment in pursuit of national goals and purposes and includes the entire astronautical capabilities of the nation." [6] This is an outstanding and *almost* perfect definition. Lupton heroically identifies three specific truths of space power: that space power is exercised through exploiting the space environment; that the aim of space power is to achieve goals and purposes; and that space power includes the entire list of space capabilities the wielder of space power possesses. However, Lupton falls short in his definition because, while his statement is a great list of space power attributes applied to nations, it is not comprehensive and general enough to be *the* definition of space power. Lupton's definition does not account for space powers other than nations. It is conceivable that space powers can be entities other than nations or states such as corporations or nonstate actors such as terrorist groups. Groups smaller than a nation can exploit space power, and this fact must be accounted for in any complete definition of space power (though for ease of discussion we will often talk of nations even though a space power is any organization which can use space activities for its own purposes). Extending this criticism, because space power exists at levels other than the national level, a nation's total astronautical capability is not universally appropriate as a metric. Although Lupton performs yeoman service to space power, his definition isn't quite complete. It is important to note that the General Theory of Space Power can be applied to any organization that operates in space. While the most familiar organization for analysis is probably a nation-state, the General Theory can be used to analyze any organization's space power, be it the human race as a whole, an alliance, a company, a government agency, or a terrorist group. All space actors can be modeled by the General Theory.

Brigadier General William "Billy" Mitchell defined air power "as the ability to do something in the air" in 1925.[7] This definition is elegant in its simplicity and forces the reader to broaden his perspective to include as air power everything that can conceivably be done in the air. Its elegance is in its simplicity, but its power is in its inclusiveness. In the author's opinion no better definition has ever been or can be presented for air power and, had future theorists not abandoned Mitchell's simplicity, great errors in space power understanding such as presented in the above definitions could have been avoided.

In the General Theory, space power is simply the ability to do something in space. Modifiers may be applied to ease analysis of space power applications at certain levels of analysis, but they cannot help but to artificially constrain the natural extent of the concept of space power if universally applied. The General Theory attempts to understand the nature of space power without any constraint of analytical level. Put simply, a general theory must be based on a general definition of its subject. Space power is the ability to do something in space. No other modifier is required.

Nonetheless, modifiers have utility. The General Theory considers definitions of space power, such as Lupton's "space power is the ability of a nation to exploit the space environment in pursuit of national goals and purposes and includes the entire astronautical capabilities of the nation,"[8] as rather a definition of applied space power—in this case space power applied to a nation-state. Therefore, the General Theory recognizes two forms of space power.

The first form of space power is its raw, basic, or general form: the ability to do something in space. The second form of space power is in its applied form—when the basic space power (ability) of an entity is used by that entity for a specific purpose. The differences in class between the basic form and applied form of space power comprise an important concept in the General

Theory which will be fully developed later. What is important is that space power in its basic form is universal among all potential space power players. It is when purposeful action employs space power (using transformers, which will be described below) that general space power is transformed into applied space power. Most space power definitions are descriptions of unique cases of applied space power (nations, states, policies, etc.) They are descriptive, important, and useful, but they jump to conclusions and cases too soon to be a valid descriptor of space power in general. Thus, the General Theory begins with ability without purpose or objective. Applied space power plays a large part in the General Theory, but is subordinate to the basic form of space power in the Logic and Grammar of Space Power.

The General Theory in Outline

Clausewitz said, "War may have a grammar of its own, but not its own logic."[9] From this historic quote we will derive a space power corollary: Space power may have a grammar of its own, but not its own logic. Therefore, the basic visual presentation of the General Theory model comprises two three-dimensional "deltas" modeled after James Holmes and Toshi Yoshihara's dual trident visual representation of Mahan's sea power theory.[10] Just as the trident was an ideal visual representation of Mahan's sea power concept, a delta shape is chosen for its ability to model important points (three points at the base extending upwards into a single point) of the General Theory of Space Power, but also for its intimate connection with the history of military space power iconography (the delta is the central figure in the shield of Air Force Space Command and is represented in many space unit patches). The first space power delta is the space power Grammar Delta, while the second is the space power Logic Delta. Both deltas are three-dimensional, allowing them to be seen from a "bird's-eye" view from the top as well as a "profile" view from the side.

The Logic of Space Power, modeled in its Logic Delta, is the warrior's art: space power used to promote the interests of the space power. The Logic of Space Power is concerned with ends and ways, in the ends/ways/means paradigm of strategy. The Grammar of Space Power, described by its Grammar Delta, is the mason's art: developing the tools with which to wield and expand space power. The Grammar of Space Power is the realm of means.

Each viewpoint signifies a different approach to viewing space power as a whole. From the top, each delta has thee outward points that link to a central point. The outside points represent three essential pieces of the grammar or logic of space power that either support, or are supported by, the central point of the delta. From the side, the delta forms a triangle with a large base extending to a single point at the top. In the Grammar Delta, the base points flow upward to ultimately support the top point signifying the goal of space power grammar development, what the General Theory describes as access. The Logic Delta, alternatively, begins at the top point—ability—and flows downward until reaching the applied base points of the delta. Each delta will be described in detail below.

While the deltas do not explain every facet of the General Theory of Space Power, they do provide a visual reference of much of the theory's content and should be considered the central facet of the theory. Although other figures will be developed and presented, the deltas encompass the critical components of the General Theory of Space Power. Thus, we can begin the General Theory of Space Power at its beginning, its grammar.

Grammar of Space Power: The Building Blocks of Space Leadership

"In these three things—production, with the necessity of exchanging products, shipping, whereby the exchange is carried on, and colonies, which facilitate and enlarge the operations of shipping and tend to protect it by multiplying points of safety—is to be found the key to much of the history, as well as the policy, of nations bordering upon the sea."—Admiral Mahan, U.S. Navy[11]

One of the tridents of sea power in Holmes and Yoshihara's interpretation of Mahan's theory (for development-inspired reasons considered the first in the General Theory of Space Power) is the grammar of sea power, which proceeds from the operational level of activity. The operational level of war is the vital link between the strategic level and tactical level of war that links national policy and strategic goals (strategic level) with the physical activity of individual troops and units on the battlefield (tactical level). Put broadly, the grammar of sea power links the logic of sea power to individual actions (or units) in producing sea power. This trident holds the Mahanian elements of sea power (slightly renamed by the authors): commerce, bases, and ships.[12] Again commerce (Mahan's original wording was "production") is the element upon which all else depends. Commerce comprises manufactured goods, natural resources, and other trade goods. Bases (original colonies) are outposts that expand access to sea lanes of communication and the availability of markets to increase trade. Lastly, ships (original shipping) are the means to transport commerce across the oceans to desired markets, and wealth to the host country. They are comprised of both the fighting navy and merchant marine of the country. Commerce is the foundation for trade, and hence the primary element of the grammar of sea power. The General Theory of Space Power's Grammar Delta is derived from this beginning.

GRAMMAR FROM THE BIRD'S EYE

The Grammar Delta of Space Power (Figure 1.1) is similar to the marine trident, but somewhat more detailed. Grammar is how space power is built and conducted by individual units to bring the Logic of Space Power to fruition through access. Grammar builds access through developing tools. It is the mason's art. Grammar is concerned with building the hardware that allows a space power to operate in space. Grammar is unique in that it is the only trident that is concerned completely with space activity. It "speaks space," while logic is concerned primarily with using space for other ends which may not necessarily be primarily concerned with space activity. Space power emerges from a foundation of the nation (or other agent's) nonspace technical, educational, and temperamental capital (the foundation of the Grammar of Space Power) and first manifests itself with the space power elements: production, shipping, and colonies. Production generates wealth from space and

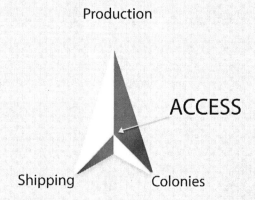

Figure 1.1 Grammar of Space Power Delta, Top View

is the backbone of economic space power. This can range from digital information to solar power to lunar resources. Colonies allow the extension of commerce to farther areas through improved access. Colonies provide markets, safe harbor for travelers, and expanded opportunities for production. Shipping plies the space lines of communication either as electromagnetic carrier waves or physical space ships to haul space products from their origin to their markets. Production, shipping, and colonies are the essential elements of space power. They span the entire spectrum of space activity, and each element must be accounted in its proper measure for a nation to be a mature and vibrant space power. These elements provide the stock of matériel necessary to fulfill the Logic of Space Power and enrich the nation through its exploitation of the space environment. These space power elements are combined (adding production, shipping, and colony elements together to form a system) in order to produce access—the capacity to perform some activity (for any reason) in a certain area of the space environment. With the elements of space power known and their ultimate purpose established—to combine the three in systems to generate access—we can turn to how that access is generated through looking at the Grammar Delta from the side.

The Profile View of the Grammar Delta

While the birds-eye-view of the Grammar Delta identified the elements of space power (production, shipping, and colonies) and *what* they are intended to produce (access), the profile view instead looks at *how* the elements *develop* access (Figure 1.2). In both the Grammar and Logic Deltas, the birds-eye-view is used for *identification* of important space power concepts while the profile view is concerned with how space power grammar and logic is *developed*. Thus, the profile view identifies a beginning and an end. In the Grammar Delta, development flows upward, from the base to the top.

The Grammar Delta begins at the foundation upon which the delta rests. This level is called the foundation and represents the principal conditions (or enablers) of space power. These conditions include that which is necessary for space power (for instance, an educated population, natural resources, industrial capacity, and others discussed in detail below) but do not directly act as space power elements themselves. For instance, engineering talent is necessary for space power development even though not all of a nation's engineers may support a space program. Proceeding upward, we find the base of the delta, or the elements of space power: shipping, production, and colonies. These are the most basic manifestations of real, developed space technology hardware (antennas, buses, rockets, and the like). These elements are mated to other ele-

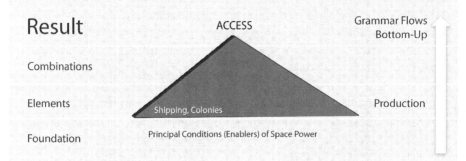

Figure 1.2 Grammar of Space Power Delta, Profile View

ments in the combinations section of the delta until a system is developed that produces the end result of these combinations: space power access, the top of the Grammar Delta. Each section of the Grammar Delta (from both bird's-eye and profile views) is an important phase of space power grammar development and will be discussed in detail from the bottom up.

Principal Conditions (Enablers) of Space Power

In the book *Space Power Theory*, Jim Oberg identifies ten items that he considered elements of space power. Identifying elements as characteristics "within a nation that makes it capable of wielding 'space power,'"[13] he names space power's elements as: facilities, technology, industry, hardware and other products, economy, populace, education, tradition and intellectual climate, geography, and exclusivity of capabilities/knowledge.[14] This is indeed a good list of important space power ingredients, but Oberg errs in calling them elements. His error is in making no distinction between the elements and enablers of space power.

Oberg's list is inspired at least in part by Admiral Mahan's six principal conditions affecting the sea power of nations: "I. Geographical Position. II. Physical Conformation, including, as connected therewith, natural productions and climate. III. Extent of Territory. IV. Number of Population. V. Character of the People. VI. Character of the Government, including therein the national institutions."[15] Here can be seen Oberg's error. He combined Mahan's two separate ideas of elements of sea power (production, shipping, colonies) and principal conditions affecting the sea power of nations into his unified elements of space power. The General Theory corrects this minor error and re-sites Oberg's elements into their proper place. Hardware and other products, as well as most facilities, are accounted for in production, shipping, and colonies and are thus true elements of space power. The remaining facilities are incorporated into industry, and the exclusivity of capabilities/knowledge is considered a subset interest of technology. Finally, a nation's resource base (such as stocks of aluminum, titanium, uranium, and other materials necessary for a space program) must be considered a foundational requirement. The remaining Oberg elements are rightly principal conditions (or enablers) of space power, which form the foundation of the Grammar of Space Power.

An incomplete list of space power enablers would include: educational infrastructure, human capital, number and character of population, natural resources, industrial base capacity, level of scientific understanding and knowledge (often incorrectly assumed to be an end of space activity in itself), economy, exclusivity of capabilities/knowledge (i.e., space industry workers versus regular industry workers as well as number of experts in space-critical technology areas), and geography. There are likely many more examples of critical space power foundational capability, but an exhaustive list of everything that contributes to a nation's capacity to build space power is likely impossible. The important realization is that anything that indirectly contributes to the national space program should be considered a space power enabler and relevant to any general theory analysis of space power.

Elements of Space Power

Like sea power, space power (in the past, present, or even in the far future) comprises three essential elements: *production, shipping,* and *colonies.* These three concepts will always be paramount, and the maturation of space power from present to visionary will only change these elements in their manifestation, not essence. Similar to what Mahan describes for the sea, space power production is goods and services that are derived from space that are traded and from which wealth is generated. Shipping is the total of services that transport space production to their respective markets as well as the lines of communication that allow the transportation to take place. Finally, colonies are the places which generate the production, give that production markets, and advance the safety of shipping by offering places of "safe haven" and protection. These elements are easy to envision for sea power. Production includes oil, other raw materials and manufactured goods that are transported over the sea. Shipping includes the many oil tankers and container ships seen throughout the world's oceans. Colonies are the many ports around the world that offer shelter in storms as well as places for shipping to pick up and deliver loads of cargo. For space power, these elements are constant even though the time frames under examination showcase very different forms of each element.

Space power elements are present no matter what form of space power is examined. Whether the aspect of space that is being exploited (a full discussion of space environmental aspects is below) is a relatively simple aspect such as high altitude or orbital dynamics using contemporary technology (as exploited by navigation, communication, and imagery satellites) or an aspect requiring relatively more advanced technology (such as extracting asteroid or lunar resources for economic use) the same elements are always present. Whether they are in use today or exist only in the lab or in a science fiction writer's imagination, all types and tools of space power can be described with these fundamental elements.

> *Production*—Space production is the total of goods and services either drawn from or transported through space. As space power matures, types of production will be added, subtracted, and expanded over time. It is also likely that the nature of the production will change dramatically from that which is prevalent today.
>
> Current Form—Near term space production mostly includes information in the form of imagery and navigation data, among others. Visual data such as satellite photos and weather data are produced by platforms in space using the unique space environment quality of high altitude. This production cannot easily be produced by other means. Though digital information does not seem to be production in a classical "goods and services" sense, it is the primary product of today's space power.
>
> Future Form—As space technology matures, future space production will take the more conventional form of physical production. Such products may include spacecraft propellants from in-situ resources, industrial quantity solar power, and microgravity-processed pharmaceuticals or structural materials.

Shipping—Shipping is the means of transporting space production from the place of its construction to its ultimate market, whether on Earth or another space destination. Like production, shipping will become more recognizable as such as space technology advances.

Current Form—Current examples of shipping are probably the most foreign looking of space power in relation to sea parallels. Most space shipping is conducted not by rockets and transports (though undoubtedly some is), but rather by electromagnetic (EM) radiation in the form of signal waves that transport essential data from a satellite to its ground station on Earth. Satellite dishes, not space shuttles, are the prime symbols of today's space shipping as they collect the transmissions that allow communication across the entire globe. Rockets are methods of shipping as well but are currently only minor players compared to antennas. They will take a larger role in the future.

Future Form—Again, future space power will paradoxically look more familiar and comfortable to the general population than its expression today. Although EM transmissions will always play an important role in space power, the future will be the era of spacecraft that are truly analogous to the merchant marine and combat fleets of today. Near term examples of future shipping are unmanned "space tugs" that can transport spacecraft from low earth orbit to geosynchronous orbit. Others will be manned spacecraft traversing the void between the Moon, Mars and Earth and later perhaps great starships traveling between solar systems with cargoes of unimaginable wonders.

Colonies—Colonies are places where the activities of space power originate and end. Colonies allow the production of space-based products, the consumption of those and other products, and facilitate the safe transportation of these goods. Colonies have the historical connotation of European villages in the New World. While space colonies are thought to be found only in science fiction, they have a very real modern day analogue in space power.

Current Form—The most common type of space power colony today is the satellite. Satellites are not the product of space power itself, though they keep the sensors and cameras where the production is made in working order. They do not transport the production data, but they facilitate its transport through providing "safe passage" to the EM signals that do. Satellites are colonies because they provide a platform for production to take place and shipping to commence.

Future Form—Again, the future form will see space power colonies approach their more traditional interpretation. Space commerce colonies will advance to become traditional inhabited colonies. Space stations, mining platforms, lunar and Martian settlements will all become vibrant areas for manufacturing goods, consuming supplies, and providing way stations for cruisers and merchant spacecraft of all types. They may also give birth to new civilizations.

As can be seen, even though today's satellites and electromagnetic information look far different from future versions of commercial empires of mining bases and merchant transport spacecraft, space power today and tomorrow share a common logic and grammar. Space commerce both now and in the far future are comprised of *production*, *shipping*, and *colonies*. Space power in any epoch can be analyzed using these terms. Now that we understand the elements, we can envision how they combine to form the Grammar of Space Power.

Combinations: Putting the Pieces Together

Above the elements section of the Grammar Delta, we find the combinations area in the center of the delta comprising much of the delta's mass. In this section, the various discrete elements (production, shipping, and colony elements) are combined to form space systems which, in turn, generate new access to the space environment. Different elements can be combined in a theoretically infinite number of ways to generate many new types of discrete accesses through new systems, and the Grammar of Space Power is primarily concerned with developing new types of space power elements and combining them in innovative ways to open as many areas of the space realm to exploitation as possible.

Space power elements can be considered like fractals in that they exist at multiple scales, they have very large and very small manifestations. Refined lunar titanium can be considered production, the space shuttle is a type of shipping, and the International Space Station is currently the closest example we have to a large scale space colony. However, every small (but complete) space system can also be visualized as a combination of each type of space power element. For instance, the GPS satellite constellation can be described as a system that combines production (the positioning, timing, and navigation information) with a shipping element (the GPS electromagnetic carrier signal) and a colony element (the GPS satellite bus itself, where production is generated and shipping originates). Combining all three elements into a functional system ultimately generates a new type of access, the ultimate aim of action along the Grammar Delta.

Access: Expanding the Playing Field

Access is the ultimate object of the Grammar Delta. Access is developed through the combination of elements into systems with each system generating a discrete new access to the space environment. Access is defined as being able to place and operate a piece of space equipment (a system of elements) in a certain area of space. For instance, by combining a production, shipping, and colony element (a camera payload, a communications link to the ground as well as a rocket to place the system in orbit, and a satellite bus) we can produce a new space power access: the capacity to take and receive space imagery from low Earth orbit.

Holmes and Yoshihara explore the all-important Mahanian concept of *access* through the dual lenses of the Logic and Grammar of Sea Power. Mahanian logic, they argue, impels governments to search out access for *commercial* reasons, and the grammar of war means upholding that access through force of arms.[16] To them, *maritime operational access* is the ability to force entry into a contested region despite military resistance.[17] Marine technology is sufficiently advanced that modern equipment can reach virtually any oceanic destination (save, perhaps, that of the deepest undersea reaches) with physical and economic ease. Therefore, maritime access is most concerned with physical safety against armed aggression. Humans have physical access to the sea easily enough that access on a physical level is simply no longer questioned. This is not so in the space environment.

In the space environment, strategic reality is opposite of that in the maritime environment. Military power capable of denying access to space is insignificant for any but the most advanced space powers, and that only with prohibitively large expense. The overwhelming amount of access denied in space is due to the limits of technology to reach it and the lack of infrastructure to support operations there, not from the actions of an active adversary. We have only very limited access to space on a physical level. Only slightly adjusting Holmes and Yoshihara's terminology, we define "access" as the ability to place a space system (combination of elements that can produce an action) in a certain area in space. Space *ability* is the capacity to conduct economic/political/military space operations in a region without significant impediment. Even though there are some differences in operational access between the space and sea environment, the underlying definition of *strategic ability* is the same: the ability to conduct unfettered operations of any type for some purpose in a given region without significant physical, economic, or military barriers. Multiple accesses combine to energize the ultimate purpose of space activity—advantage. This advantage of space activity is expressed through the Logic of Space Power.

Logic of Space Power: The Ultimate Strategic Purpose

Military precautions, and the conditions upon which they rest ... while they have their own great and peremptory importance, cannot in our day, from the point of view of instructed statesmanship, office-holding or other, be considered as primary. War has ceased to be the natural, or even normal, condition of nations, and military considerations are simply accessory and subordinate to the other great interests, economic and commercial, which they assure and so subserve.... [T]he starting point and foundation is the necessity to secure commerce, by political measures conducive to military, or naval, strength. This order is that of actual relative importance to the nation of the three elements—commercial, political, military.—Rear Admiral Alfred Thayer Mahan, U.S. Navy[18]

Mahan is known as a great military thinker, but in his model peaceful commerce is the pivotal foundation of sea power. As Mahan says, there is a hierarchy in the types, as well as different levels from which we can understand sea power—such as it is with space power.

James Holmes and Toshi Yoshihara explore the levels of sea power in Clausewitzian terms. In their 2009 article *Mahan's Lingering Ghost* they also describe the logic of sea power.[19] The logic of sea power is the national strategic character of sea power—the description of sea power's utility to a nation. Holmes and Yoshihara fashion the graphical representation of the logic of sea power as a trident, with the top spire commercial, and the flanking spires political and military.[20] These three points of the trident mark the three primary uses of sea power. Commercial sea power provides the nation with wealth and prosperity through trade and production. Political sea power provides the wielder the means to exert control over other nations through diplomacy and granting or restricting access to the sea lanes of communication under its control. Finally, military sea power can both defend the nation and its commerce as well as project power to drive an enemy's flag from the waters and deny an adversary acess to the sea.

Access to the sources of economic well-being—foreign trade, commerce, and national resources—ranks first within the logic trident, and military might third.[21] The Logic of Sea Power is this: sea power's ultimate goal is to generate wealth from the sea. Commerce is the true path to both affluence and national greatness.[22] Commercial sea power provides this wealth and seeks more wealth through access to markets. Political sea power ensures and expands commercial access (the critical impact of access will be discussed later) through diplomatic pressure and other nonviolent means. Military sea power is the sword and shield of the nation, as well as the guardian of sea commerce and defender of its access to the sea.

The Logic of Space Power is a sibling to that of sea power. Space power's ultimate purpose is to generate wealth from space activities, and commerce is the true path to national greatness in space. Instead of a trident we again use a three-dimensional delta (a common symbol in Air Force Space Command) to express the Logic of Space Power visually. To fully understand the Logic Delta, we must look at it from a birds-eye-view as well as a profile view. From a birds-eye-view, the points are the same as the two-dimensional sea trident: the center and tip of the delta is commerce, and the two supporting flanks are again political and military. The delta's top point, or spire, is the ability to act in space—the ultimate definition of space power.

LOGIC FROM THE BIRD'S EYE

Looking at the Logic Delta from the top (Figure 1.3), the three spires represent economic, political, and military power, with the central apex representing ability. The spires represent applied power in service to some nation's interest. The apex, ability, is the raw source of capability to act in space from which all applied space power (economic, political, or military) flows. Ability

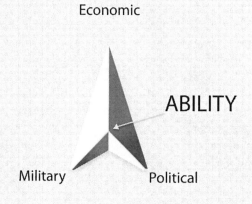

Figure 1.3 Logic of Space Power Delta, Top View

is pure space power, the ultimate expression of the work that the Grammar Delta accomplishes.

Space power, in its most elemental form, is wealth from space whether scientific, philosophic or—of greatest importance—material in nature. Economic space power produces the all-important wealth from space, and thus is given primacy of place at the leading edge of the Logic Delta. Political space power generates and leverages diplomatic and other nonviolent elements of national power to expand and ensure commerce's access to markets and the sources of wealth. Lastly, military space power defends commerce's access to its essential markets and areas from which they produce wealth, and denies adversaries access to space wealth in times of conflict. However, commerce is always the paramount end and primary means of generating space power. Political power and military space forces do not exist as ends unto themselves, but are merely the means to expanding and ensuring space commerce: the vital lifeblood of the healthy space power nation.

THE PROFILE VIEW OF THE LOGIC DELTA

Just as in the Grammar Delta, the profile view of the Logic Delta (Figure 1.4) is concerned with the development of the Logic of Space Power. However, instead of flowing from the bottom up like grammar development, Logic Delta development flows from the top down. At the very top exists pure space power, the ability to do something in space. This represents the combined sum of accesses derived from the Grammar Delta. Below the raw capability of ability, we have the body of the Logic Delta, the transformers. Transformers are the ideas and concepts used to translate raw ability to do something in space into concrete applied power. Transformers turn the capacity to operate in space into concrete power from space that can be applied to economic, political, or military use to achieve national objectives. At the bottom of the Logic Delta is space capability applied to the national interest, where applied economic, political, and military power from space is generated and wielded.

Ability: Space Power in Its Rawest Form

Because space power is nothing more than the ability to do something in space, increasing the ability to do anything in space is the foremost interest of a maturing space power. Development of pure space power is the exclusive realm of the Grammar Delta, generating new access. New discrete accesses are created by every successful combination. Ability is simply the integration of

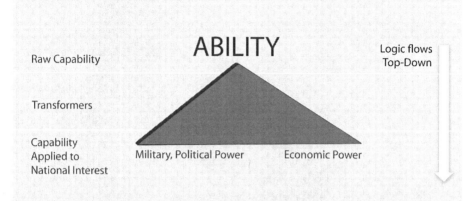

Figure 1.4 Logic of Space Power Delta, Profile View

all of a space power's discrete accesses (the capabilities of each of its space systems) into a comprehensive whole: the nation's (or agent's, because any organization can be space power) entire ability to do anything in space.

Pure space power exists in a nation's space power ability. Ability is a measure of nonapplied, but immediately realizable, potential at any given time. Ability in itself has no purpose or direction—it is the capability the nation has to act in space for any reason. Because it has no inherent purpose or direction, ability requires operational concepts and purposeful action to create applied space power. These ideas and actions transform ability into applied power as we proceed lower on the Logic Delta.

Transformers: Ability to Application

Raw space power is the ability to do something in space. Only by using this raw space power in pursuit of some interest (be it national interest in the case of a nation, or corporate interest if the space power agent is a company) is it ever truly useful. Space power only becomes useful by turning it into agent (i.e., national) power. Raw space power (ability) flows downward through the

Logic Delta and reaches the transformers section of the delta. It is in the transformers section that the raw space power turns into applied space power: economic, political, or military power. It is through transformers that this raw
power becomes applied power.

Transformers are the concepts that turn space power ability into applied
power, the ideas for making space power useful to national objectives. The
General Theory identifies three types of power (economic, political, and military) that space power can be transformed into productive use. While the
transformer is essentially an idea to put a space power ability to a useful purpose, transformers take three different primary shapes depending on which
type of power is being produced by space activity. Thus, transformers generally
take the form of a business plan, soft power concept, or military doctrine.
Effective and sustainable transformers are critical for the development of space
power as raw ability can be transformed into an almost infinite variety of purposeful uses. Once a space power ability is established, it needs only successful
ideas to turn it into applied and purposeful activity beneficial to the nation,
and the more successful transformers a nation possesses, the more applied
power can be generated from space power.

ECONOMIC SPACE POWER

Space power, like sea power, is primarily economic power, and thus takes
its place as the leading spire of the Logic Delta. Economic power's end is the
generation of wealth and accomplishes this through the development and maturation of business entities, usually businesses. Therefore, the primary transformer to generate economic space power is the business plan. A business plan
can be defined as "a document that summarizes the operational and financial
objectives of a business and contains the detailed plans and budgets showing
how the objectives are to be realized."[23] Using the General Theory framework,
a business plan is a detailed plan to use space power ability to generate wealth,
the ultimate "operational and financial" objective of any business. The business
plan takes a raw space power ability and puts it to use in a profitable and
sustainable process that generates wealth—economic space power. More successful business plans produce more successful space businesses and more economic applications of a nation's raw space power—ultimately increasing the
nation's wealth from space. Innovation in space business plans will result in
massive growth of wealth from space.

As an example, there are currently many successful communications satellite–based business plans as exemplified by the multiple successful space communications companies in existence. Not only do communications satellite

businesses exist for telephone and internet connectivity, but also satellite television services. From the communications satellite, business plans have been developed that have spawned multiple industries, not simply multiple companies. However, although the technology exists to reach the nearest asteroid, multiple engineering concepts exist to mine asteroids, and there are identified resources potentially reaching into the trillions of dollars (i.e., the production, shipping, and colony elements necessary for this operation—the requisite grammar—exists), no profitable asteroid mining company is currently in operation. No profitable business plan has ever been developed and proven to work (though James Cameron's Planetary Resources, Inc., plans to give it a shot). Asteroid mining may be considered science fiction by many, but if a successful business plan based on this space power ability is ever developed there is no doubt that a revolutionary new form of economic space power will have been demonstrated. Therefore, business plans are an essential—perhaps the most essential—type of transformer in space power theory. However, it is not the only type of transformer that can exist.

POLITICAL SPACE POWER

Political space power is raw space power ability applied to political or diplomatic ends. Political space power can be coercive (such as power from intelligence activities) or it can be, to use Joseph Nye's term, co-opting and attractive (the so-called soft power approach). However, since space-based intelligence has long been intimately associated with military space power, political space power in the General Theory will focus on the soft power component. Joan Johnson-Freese says that the Apollo 11 moon landing was "perhaps [the United States'] most inspiring global moment" and noted that it "symbolized U.S. leadership and its future-oriented direction."[24] This space power event no doubt caused many nonaligned nations to favor the United States during the Cold War, a critical example of political space power. Joseph Nye called this type of power "soft power." Soft power is "getting others to want the same outcomes you want" and rests on the ability to shape the preferences of others."[25] There are many examples of space power–based soft power operations. The Apollo moon landing may be the best known. Another clear example is the reconfiguring of the Reagan-era Space Station Freedom program into the International Space Station for diplomatic ends.

Although coercive forms of political space power are possible, they often are political uses of different types of applied space power. For instance, space-based intelligence coercion leverages military space-power intelligence concepts, and potential sanctions against an offending nation using economic space power can be envisioned. However, it appears that political space power is often the province of soft power owing much to the awe-inspiring and idealistic views that many people place on space activity. Therefore, the General Theory places soft power concepts as the foremost example of the political space-power transformer.

MILITARY SPACE POWER

Military space power is using raw space power for military missions. Military space power can be combat power as well as military operations other than war through or from the space medium. In the General Theory, transformers are the means to convert raw power into applied power. According to Air Force Major General I.B. Holley, the transformer role in converting ability to operate in a medium to achieving power through action in the medium is doctrine because doctrine is inherently concerned with *means*.[26]

While any military concept can act as a transformer, fully developed doctrine is the preferred transformer for military space power. Why? As General Holley explains:

> Doctrine, as officially promulgated, has two main purposes. First, it provides guidance to decision makers and those who develop plans and policies, offering suggestions about how to proceed in a given situation on the basis of a body of past experience in similar contexts distilled down to concise and readily accessible doctrinal statements. Second, formal doctrines provide common bases of thought and common ways of handling problems, tactical or otherwise, which may arise.[27]

Doctrine is a set of military concepts that have proven successful in the past. Doctrine is, then, similar to a proven successful business plan in that it has proven successful in the past and will probably remain successful in the future. Like business plans, however, useful doctrine has a shelf life and must continually adapt to new circumstances. Therefore, military space doctrine must always be a constant source of study and experimentation by space forces in order to have the best means to exploit their ability to operate in the space environment.

The experimentation, testing, and validating of new concepts such as business plans, soft power applications, and military doctrine is a key avenue in the development of space power. While physical tools are a primary source of development in the Grammar of Space Power, space power development through logic is primarily driven by developing new transformers for existing tools. The transformer is a central concept of the General Theory.

Applied Space Power

Applied space power takes the form of economic, political, and military power, the three concrete manifestations of power. It is important to note that at the applied level, national power is power and not strictly considered as space power. It is power in the interagency and joint sense; it is national economic, political, and military power and the ultimate interest of the nation is on the effects of that power, not the fact that it is generated from space. National leaders are interested in power effects at the applied level. It is only space leaders in charge of space activity who are interested that certain capabilities are derived from space systems.

Readers will no doubt draw parallels between the triumvirate model of applied space power (economic, political, and military) with the well-known military acronym DIME (diplomatic, informational, military, and economic) power. Both attempt to model the same dynamics. The General Theory keeps applied power to three manifestations for aesthetic reasons (a delta has only thee base points) but mostly in order to stay true to Mahan's original three-pronged description of sea power. Indeed, a strong case can be made that both diplomatic and informational power (the DI in DIME) are simply different mechanisms of political power. As Clausewitz said, war is the continuation of politics by other means. Diplomacy and information power comprise the normal operations of politics by *regular* means.

Linking Logic and Grammar

ABILITY VERSUS ACCESS

Of critical importance to the General Theory is the link between space power's logic and its grammar. To define the link between space power's logic and grammar we must clearly delineate the concepts of *ability* and *access*. While closely related, they are not synonymous terms.

Space power's foundation is built through the Grammar Delta. Space power's elements (production, shipping, and colonies) are combined in various ways to produce *access* to space. *Access* is again defined as the capacity for an entity wielding space power to place a space power element (of any type) in a specific area of space for some useful purpose (whatever it is). For instance, if a robotic spacecraft can be sent to Jupiter it can be said that the probe's owner has access to Jupiter with said probe. Since exploration spacecraft have been sent to Jupiter, the United States has access to Jupiter with robotic spacecraft.

However, since no manned spacecraft yet designed (much less built) can be sent to Jupiter for lack of sufficient engines and environmental equipment, the United States does not yet have manned access to Jupiter. Note that access does not require or manifest a *purpose*. Access is merely a statement of potential; it does not need any justification or rationale behind it. Access is also a discrete term. We have access to a specific place with a specific element. As exemplified above, we can have access to a place with one specific element or type of element but perhaps not others. When we begin to add conscious purpose to a mission we have the access to achieve, and when we aggregate our different accesses, we begin to develop *ability*.

Ability is the pinnacle of the Logic Delta. Whereas in the Grammar Delta we build from the space power elements to access, the Logic Delta begins with ability and flows down through the transformers to translate that space power ability into specific space power abilities that add to economic, military and political power. But the beginning of the Logic Delta is *ability*, space power in its rawest and most primitive form. And ability is formed through the Grammar Delta's access.

Ability is the sum total of all discrete accesses plus the all-important *knowledge* of how those accesses can be used for benefit and *intent* to use access for some *purpose*. Since basic access is a discrete term (do we have access to place A with element B?), the sum total of things we can do in space is necessarily the summation of all these discrete accesses. Of course, there is no need to specifically list all accesses of elements to all places in space, the sum total of access can generally be guessed and aggregated easily. However, the aggregate access term is not the entirety of ability. The critical difference between logic and grammar is *intent to use*. Grammar builds the tools, logic informs if and how the tools will be used to produce some effect.

Refer back to the definition of space power: the ability to do something in space. Just because we have *access* to an area does not mean we have the *ability* to manipulate the area for our own purposes. Just because we have the capacity to physically do something, if it never occurs to us to do it (or we have no interest in doing it at all), can we truly say we have the ability to do it? Since ability is really raw ability (access with the intent to use, though no specific intent as yet) because space power in its general form is only raw ability, intent is nothing more than the *spark* that links access to a desire to use that access for our benefit. Ability is space power, but in order to be truly wielded, that space power must be applied (through the transformer mechanisms) into applied power in the economic, political, and military sphere. Thus, as soon as ability is established, it flows down the Logic Delta into specific space efforts that enhance the space power entity's economic, military, and political power.

Thus, the Grammar Delta (the building blocks of space power) and the Logic Delta (the intent and application apparatus of space power) are linked through the all-important concepts of access and ability. Access is the capacity to place an element in a specific area in space, and the sum total of discrete accesses available to the space power entity in question plus the intent to use that aggregate access for any purpose is ability. Access and ability connect the Grammar and Logic Deltas to form a complete and collimated model of space power.

THE FEEDBACK LOOPS

Try as we might to separate logic and grammar into their own contained spaces, it is an unfortunate (for simplicity of the model) truth that there are feedback loops between the Grammar and Logic Deltas: grammar builds that which logic finds useful (ability through access), but logic also informs the selection and development of grammar's space power foundation inputs (specifically, through the transformers).

Strategic Access: Current Limitations on Ability

It is ability (the aggregate of accesses plus intent to use), or total access integrated at the logic level, that is most important to space power. However, even if technology exists to reach a certain area in space, access to that area may still be blocked by the purposeful activity of an adversary. Thus, while we may have access to an area, we may not have strategic access to an area due to some limitation placed on us by an adversary. At the operational grammar level of space power, access can take different hues based on whether one's access is limited from a commercial, political, or military point of view (i.e., can we reach a part of space physically, are we forbidden by treaty to operate there, or is an adversary keeping us out of the area by force?). Thus, strategic access is bounded by the most restrictive operational grammar access. In today's sea realm, strategic access is often bounded by military force (an enemy fleet blocking access to a port, not our lack of ships able to reach that port). In the space realm, access is currently bounded by technology limitations (we have no rocket capable of taking one thousand people to Mars). Since there is little limit on access through strategic means, in space technological access normally means we have strategic access as well. Therefore, the best way to extend strategic access in space is not to invest in ways to obstruct an enemy's ability to limit our access to space through military means and obsess over "space as a

contested environment" as most military space leaders continue to do in prepa-
ration for war, but to invest in technology and attack the greatest current
enemy to space access—the space environment itself—by investing in technology to open economic access to more of space for all human endeavor, primarily commercial development.

Even though we should use propulsion technology to expand access, there
is still a large military role in expanding strategic access. Expanding access to
commerce will by default extend access to military operations, and the military
must always fulfill its primary function: to protect friendly commerce and to
deny access to hostile commerce during times of war. But how do we expand
strategic access in peacetime? Through the *peaceful strategic offensive*. Expanding commerce as far out into space as possible doesn't just increase wealth, it
also expands access and ability to operate, advantages that can be utilized by
any applied type of space power: political and military, as well as commercial.
Therefore, the peaceful strategic offensive—expanding commerce—yields
political and military benefits and national advantage that need not be procured through violence. Commercial activity is every bit as strategic in national
power as military or diplomatic coercion, and is oftentimes far more palatable
to the international community. This is only one reason, of many, that commerce takes priority in the development of the Logic of Space Power.

Levels of Space Exploitation: A Hierarchy

The space power elements (production, shipping, and colonies) are common throughout all space operations for all conceivable times past, present,
and foreseeable future. However, the Mahanian space power model understands that these elements can be combined in many different ways to produce
profoundly different forms of space power. There can be military, commercial,
and political forms of space power (breadth across the potential uses of space's
environmental characteristics), but different forms can also be seen depending
on *what* environmental characteristic of space is being exploited. These different characteristics not only describe what can be gained through space
power but offer a characterization of the maturity and potential reward regarding the space activities of the spacefaring nation being examined through which
characteristics it is primarily exploiting for benefit. To recognize this facet of
space power, the model identifies six critical space environment characteristics
and lists them in ascending order (based on operational maturity expressed
through required strategic access and potential economic reward) that can be
used to classify the maturity of the space power nation in question.

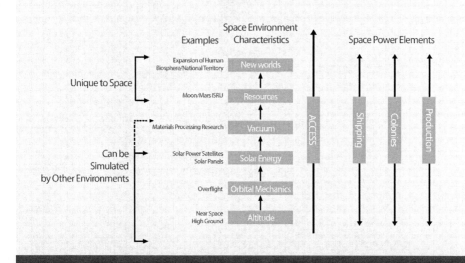

Figure 1.5 Space Power Hierarchy of Access

1. *Altitude*—This is the most basic environmental characteristic of the space environment and easiest to exploit. The first payloads were cameras or other instruments that took advantage of the great heights reached by the first suborbital rockets, and today the altitude advantages of space are exploited in almost every profitable space activity. Imagery satellites use the altitude advantage of space travel to take wide photographs for civil, commercial, or military use. Communications satellites use altitude to act as signals repeaters.

   Though altitude is very valuable and easy to exploit, a robust space power cannot be based upon it alone. Altitude can be exploited by equipment other than space systems, such as aircraft. Indeed, military space circles debate whether high altitude "near space" balloons should be considered space forces. From this space power model's perspective they should not. The altitude advantage is a weak form of space power because it can be easily replicated by other environments (high altitude airborne vehicles) and requires very little strategic access to space for a nation to exploit it. Indeed, altitude does not even require access to orbit in order to exploit. Thus, altitude is the most basic and least powerful environmental characteristic of space to exploit.

2. *Orbital Mechanics*—Orbital mechanics, such as the ability of objects to stay in orbit for long periods of time without significant maintenance requirements and the ability to overfly swaths of earth with regularity, is an environmental characteristic more difficult to exploit than altitude, but is still a very basic application of space power. The first artificial satellite, *Sputnik*, proved the overflight concept and demonstrated the utility of exploiting orbital mechanics for space power. When combined with exploiting altitude, we begin to see the rise of imagery and commercial satellites that can stay on station for years. However, orbital mechanics was exploited even earlier than space altitude exploitation in the form of ballistic missiles, which use suborbital trajectories to reach their targets.

Combining the space environment's altitude and orbital mechanics, along with rudimentary solar energy considerations, we enter the modern realm of space power as satellite constellations. This is a potent combination, and most space power theories have focused at this level. Achieving orbit is a large leap for strategic access over simple suborbital sounding rockets. However, it is still a relatively minor application when compared to the total possible activity in space. Orbital mechanics make persistent presence easier than alternate methods, but not impossible. Therefore, exploiting orbits does not make any activity genuinely unique to space and stopping at this level makes for an immature command of space.

3. *Solar Energy*—The abundance of solar energy in near Earth space is a critical environmental characteristic for space power. With the previous two lesser characteristics, it makes today's modern constellations possible with solar arrays for providing ready power to satellites. It is a higher order of environmental characteristic than altitude or orbital mechanics because it presupposes the first two in order to be exploited. Solar power is generally only needed for longer missions made possible by orbital mechanics. Heavy industrial exploitation of this characteristic in the form of solar power generation satellites to beam power to Earth is a potential future application with great economic potential.

The long-term missions that make solar energy exploitation necessary and the ability it brings to extend mission length allows exploiting nations to greatly multiply their strategic access to space. However, this characteristic can be exploited by earthbound technologies, albeit at a lower efficiency. Therefore, since the three characteristics exploited by modern space power (altitude, orbital mechanics, and solar energy) can be simulated by other environments, there are sometimes calls to retreat from the space power for other opportunities, as recent plans to find alternatives to GPS satellite navigation for the Air Force attest. This provides reasons to believe that space power using these three characteristics may not be sustainable or stable in the long run.

4. *Vacuum*—Space is near total vacuum. Combined with microgravity (technically an extension of orbital mechanics), space itself offers a unique environment to conduct research and manufacture advanced materials of high purity (exploiting vacuum) and advanced uniform structure (exploiting microgravity). Exploiting vacuum in space for industrial purposes may offer a leap that increases commercial space power by orders of magnitude.

To be sure, vacuum and microgravity can be generated on Earth (so the model assumes that this characteristic is still not unique to space), but exploiting space may be the only way to make this type of manufacturing economically viable. Exploiting vacuum greatly expands strategic access because it implies that the space infrastructure supporting the industrial effort is sufficiently large so that firms cannot only get to the area of space in which they are operating, but are also able to perform complex activities necessary for manufacturing—truly exhibiting freedom of physical and economic action in space. This is the ultimate form of strategic access in an area: the ability to perform complex economic activities necessary to fully exploit the advantages of the area of operations. Using vacuum is the first type of environmental characteristic that truly offers spacefaring nations a critical capability that may be permanently restricted economically without exploiting space itself.

5. *Resources*—The space beyond Earth contains many raw materials and resources. The Helium–3 isotope present in small quantities on the Moon and in large quantities in the gaseous outer planets may be the critical component for economic and clean fusion power reactors. Water on the Moon could fill a critical need for future space operations. Metallic and volatile chemicals located on asteroids and planetary bodies truly add to the resource base available to nations and humanity as a whole. Their presence necessitates their inclusion in any space power model in order to even pretend completeness.

Space-based resources can be used to defray mission costs such as lifting large amounts of construction materials from Earth to build a lunar base or simply processing a new tank of gas on a Mars sample return mission. Extracting and refining space-based physical resources for human use is the second most advanced level a space power can achieve. Resource extraction requires heavy industrial activity and a large degree of strategic access in order to carry out. Adding potentially large gravity wells when dealing with planets and resource exploitation demands a highly sophisticated space infrastructure and a near total strategic access to the area. However, there is one last superior quality of space for a mature space power.

6. *New Worlds*—Processing space-based physical resources may be a sophisticated expression of economic activity in space, but they can be exploited through remote mining platforms or outposts with little or no human presence. A complete and systematic exploitation of all space environment characteristics requires the acceptance of space as a human environment. Instead of seeing the Moon only as a visiting stop, a place for scientific research or to build mines, exploiting new worlds requires the spacefaring society to use space locations to extend human life and culture. In short, it required colonization of the solar system and space beyond.

Instead of merely gaining earthbound advantage, or adding new resources into the national economy, space under new worlds allows a true extension of the nation's polity off-planet—or the possibility of creating of an entirely new polity altogether. The completely developed mature space power must be able to spread its civilization off Earth and to the areas of space it controls. In colonized areas, strategic access becomes complete. Although advances in technology could continue to make operations in the colony easier, all human activities are able to be conducted in the colonial area. Through exploiting space areas as truly new worlds, the space power nation turns dead worlds into new human environments.

Thus, the model's six space environmental characteristics and their impact on strategic access form a holistic continuum that brings us from small suborbital rockets to satellites to the vast, futuristic colonial empires of science fiction. All are examples of space power, and all are built through the elements of space power. With the model now complete, let us examine some possible lessons we can derive from it.

Developing Space Power

Economist Joseph Schumpeter's theory of economic development can be used as a proper model of space power development because Schumpeter's model shares the same characteristics as space power itself. Indeed, isn't space power development just a form of economic development through space activity? Schumpeter argued that economic development has three salient properties: it comes from *within* the economic system itself and is not a reaction to outside information, it occurs *discontinuously* and is not a smooth process, and it brings *revolutions* which fundamentally change the status quo and result in new equilibrium states.[28] Development of space power is essentially the same. Space power development is an *endogenous* process that takes place within the Logic and Grammar Deltas of space power and is not something that just "happens" from activity outside the model. Space power development is also not a smooth process, but occurs in fits and spurts as new manifestations of the elements are combined in different new ways to develop new avenues of access and new capabilities expand the ability to exploit the space environment. Lastly, developments in space power often revolutionize the grammar (but not the logic) of space power by expanding space power to new vistas which fundamentally transform the balance of space power and what space forces look like (satellites to space stations to human colonies, etc.). Because space power develops in ways remarkably similar to Schumpeter's concept of economic development, space development can be described in Schumpeterian terms.

Schumpeter says "Development ... is then defined by the carrying out of new combinations" of productive elements.[29] In space power development, the productive elements being combined in new ways are the space power elements—production, shipping, and colonies. Schumpeter's concept of development (and the General Theory's concept of development of space power) covers five specific cases (which we will call paths):

1. The introduction of a new good (for our purposes a new type of element: production, shipping, colonies) or of a new quality of good.

2. The introduction of a new method of production not yet introduced to the industry in question (but not necessarily a new discovery in and of itself) that can exist in a new way of handling a commodity (space power element) commercially.

3. The opening of a new market, whether it is new or an established one if the host nation's manufacture has just achieved access.

4. The conquest of a new source of supply whether it is new or simply newly available to the host in question.

5. The carrying out of the new organization of any industry.[30]

These five cases of development need not be significantly altered to serve as the five paths of space power development, but it is good to explain what activities under each path would look like in a space power perspective. Space power is developed through the paths in the following ways.

PATH 1

The introduction of a new good in our model is a new element of space power (production, shipping, or colony), and this path consequently interfaces with the model at the element level of the Grammar of Space Power. The elements form the base of the Grammar Delta and are the most visible examples of space power. Path 1 may be the easiest development path to visualize. This type of development is maturing space power through the creation of better physical tools. Developing a newer and more powerful engine that, in turn, creates a more powerful spacecraft (example of shipping) is an example of Path 1 development. Anything that is a new element of production, shipping, or colony can be considered an example of Path 1 development (see Figure 1.6).

Path 1 development is achieved primarily through space-centric research and development (R & D) in various private and governmental labs. Path 1 is space power development through *hardware*. Higher quality space power elements inherently produce greater access because they are inherently more durable, flexible, and of higher performance. If they aren't these things, the elements aren't higher quality than those that came before! However, not only must new elements be developed in a lab, they must also be produced and fielded to increase space access. Therefore, Path 1 development can also be spurred by increasing the prototyping capability of the industrial base along with other ways to increase the speed of new production (not simply greater production) in the factories relevant to space power. Tightening and increasing the speed of the new technologies production and incremental design improvement and production cycle is an important facet of Path 1 development.

Path 1 development will always be critical because the elements of space power are the basic building blocks of space power and may be considered the most important pieces of space power simply because they are the hardware and tools with which to act. Without the elements of space power, an entity cannot be a space power at all. While other items in the General Theory of

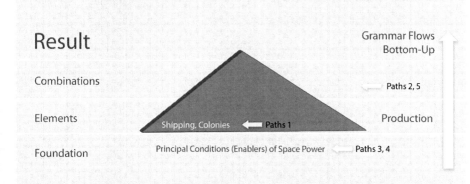

Figure 1.6 Grammar Development Paths

Space Power are critical to space power, in a large sense they are simply modifiers to the elements. Because the elements of space power are so important, Path 1 development will always be a vital development path of supreme importance to a responsible space power.

PATH 2

Path 2 development is the introduction of a new method of production not yet introduced to the industry in question (but not necessarily a new discovery in and of itself) and can exist in a new way of handling a commodity (space power element) that improves access to space. Path 2 is, in economic parlance, space power development through technology, knowledge or innovation spillovers from industries not necessarily related to space power in and of themselves. Economists have long been aware that a great driver of innovation was cross-industrial pollination of new ideas. This pollination is a key theory for why inter-industry agglomeration (i.e., many different industries existing in one major metropolitan area such as New York) seems to generate a larger amount of innovation for all industries in a major city. A common scenario is the "bar napkin event" where two people from different industries work out a new breakthrough process after hours at a bar by applying the process from one industry to the other, creating a transformational innovation. This advantageous innovative event is known in economics as a knowledge spillover.

Knowledge spillovers often result from adapting the processes of one industry into the methods of another. There is really no new breakthrough necessarily created, but important (and often industrial breakthroughs) advances can be made through these knowledge transfers between industries. As Schumpeter says, knowledge spillovers can take the form of new methods of production or new ways of handling a commodity (space power element) in the industry which received it. Therefore, Path 2 innovation can affect two areas of the Grammar and Logic of Space Power (see Figure 1.7).

The first way Path 2 development can impact the Grammar and Logic of Space Power is at the combinations level of the Grammar Delta. Space power access is created by combining the space power elements together in new and different ways (the process of development) that form new capacities for elements to operate in new areas effectively. A new way of handling an element inherently creates new ways with which to apply the affected element to achieve new access capabilities.

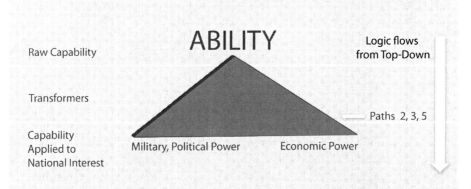

Figure 1.7 Logic Development Paths

The second place that Path 2 development may affect space power is at the transformer level of the Logic of Space Power. Transformers use economic, military, and diplomatic concepts to translate raw space power into national economic, military, and political power. New ways of handling space power elements may also translate into new ways of using the element for national power purposes. Therefore, Path 2 development may affect both combinations made using the space power elements and how those new combinations (or more specifically, the new accesses and abilities created by those new combinations) can be used to generate national power.

Since Path 2 development is generated through innovation by knowledge spillovers between industries, it can be stimulated by ensuring that a portion of the space power development industry is in close contact with other industries to keep an open line of communication available. Path 2 development, then, makes all industries at least tangentially important to space power development, not simply the industries directly related to space power. For it is unknown in advance which age-old process from which random industry will prove to be the key to creating an essential Path 2 development breakthrough for space power. Therefore, it is essential that those charged with space power development in any entity be aware of developments in other seemingly nonrelated fields. Space power development is a necessarily multidisciplinary field in many ways, and Path 2 development is a large reason why this is the case.

Path 3

The opening of a new market, whether it is new or an established one if the host nation's manufacture has just achieved access. Because space power is primarily economic (military and diplomatic power can only be maintained as long as money continues to flow), Path 3 development is vastly important to any space power entity. New markets open new avenues to generate wealth using space power. Established markets new to the space power give the newly entered space power a new method of generating wealth. Completely new markets offer the space power that initiated for the market a superb opportunity to capture that "blue ocean" (black ocean?) market for itself, greatly increasing its wealth potential.

As with Path 2 development, Path 3 development impacts both the Grammar and Logic of Space Power. The first impact of Path 3 development is to the foundation level of the Grammar Delta. New access to an existing market allows the space power that achieved this new access to incorporate the market in question into the economy foundation of the Grammar Delta.

Any access to a market allows the exploitation of that market to help fund a space power development program, as well as produce revenue for other essential activities. This is perhaps the easiest way to initiate Path 3 development, through competing in new markets that have already been established. From a strategic perspective, entering a new market will impact competitors by both threatening established revenue streams (not entirely described by market share) and making them defend their presence in the market through heightened competition. Path 3 development through entering new space markets is a very valuable development path for both absolute (more money) and relative (damage competitors) strategic reasons.

The second type of Path 3 development, the creation of a new market altogether, comes into play at the transformer stage of the Logic Delta, specifically connected to new economic and business methods through transformational ideas and methods to generate economic space power. Economic transformers by their very definition create new markets enabled by space power. Any time a profitable new product or service can be provided by combining the elements of space power, an economic transformer has generated a new market. These new markets are critical enablers of generating abnormally large amounts of revenue through economic space power. As will be discussed in a later section, monopolizing as much as possible these new space-enabled industries is critical to the economic health of the space power in question and are among the most valuable manifestations of space power possible. Path 3 development through existing markets is valuable. Path 3 development through new markets is an economic national power multiplier of almost absolute importance.

Path 3 development is encouraged through the inculcation of an entrepreneurial spirit into the entire space power enterprise. Although military and political space power can be critical enablers and defenders of this type of development, Path 3 development is almost entirely the responsibility of the economic and business sectors. The space entrepreneur, more so than perhaps even the military space professional, is the critical asset to space development and space power. Encouraging a large, innovating, and risk-taking space business culture mostly free of government interference (except where national security and interests demand limitations on behavior) is perhaps the most important thing a government can do to mature its space power. Economic power is the cornerstone on which all other national powers rest, and space is no exception. This fact makes Path 3 development among the most sought avenues to develop space power.

PATH 4

The conquest of a new source of supply whether it is new or simply newly available to the host in question is the Path 4 approach to space power development. Path 4 development results from the acquisition of a new supply of strategic resources necessary for the maintenance of space power. These resources can either be rare or base resources. These resources can be as common as iron or aluminum or as rare as antimatter—if they are necessary for the maintenance of space power, they are useful.

Path 4 development impacts the foundational stage of the Grammar Delta, specifically the Resource Base space power enabler. This is perhaps the most straightforward development path there is. With a larger resource base, an entity's space power will have a larger supply of raw material with which to build and maintain its space power. Some resources may be more critical than others. Fissile or fusion material will likely be more important and rare to space power than iron ore if nuclear rocketry ever becomes commonplace in space operations. Nonetheless, accumulating any new supply of a resource that is necessary for space power to function can be considered Path 4 development.

Path 4 development can be stimulated in a number of ways. First, resource supplies can be accumulated to increasing access to new deposits and developing the ability to exploit them. This exploitation can occur either terrestrially or in space. Given the increasing evidence of large amounts of resources (including water, metals, volatile chemicals, and even sunlight) available in even near–Earth space, space power will likely depend on space resources at some point in the foreseeable future. Indeed, the most important near-term space-power revolution will likely be widespread use of space resources in developing space power elements such as production (in the form of raw material goods) and colonies (raw materials for the construction of new space outposts).

A brief discussion of the space power element of colonies in relation to Path 3 and Path 4 development is in order. Space colonies (in their classic science fiction form as habitable cities and outposts off Earth) will be important enablers in Paths 3 and 4 development as soon as human spaceflight becomes regular and common. New human outposts off Earth will become markets for important items almost immediately. Oxygen, water, and food (among many other basic necessities) will become traded commodities immediately upon the founding of even a small manned outpost. NASA's commercial cargo program to the International Space Station is a current example of new markets (Path 3 development) that spring from new colonial requirements. Therefore, an important catalyst for Path 3 development will necessarily be the foundation and expansion of new colonies.

Likewise, colonies will be particularly important elements in Path 4 development through extracting resources from space deposits. Just as many towns and outposts of the frontier American West normally got their start as mining camps, so will space colonies likely develop from economic activity related to mining valuable deposits of space resources. In order to gain access to space resources and develop the ability to exploit them, human colonies of some size will probably be necessary to supply needed labor and/or expertise that robots may not be able to provide either physically or economically. Thus, it is likely that Paths 3 and 4 development will both spur the creation of and be dependent on the establishment of human colonies off Earth.

PATH 5

The carrying out of the new organization of any space industry is considered Path 5 development. This is perhaps the least appreciated path to development. Schumpeter, as an economist, considered the organization of an industry in economic terms. If an industry went from an environment of competition to an oligopoly or monopoly structure, or vice versa, this was an inherent development in the organization. This type of economic development is perfectly viable as a space power development path. For instance, the rise of "New Space" companies such as Space Exploration Technologies (SpaceX) or Bigelow Aerospace to break the oligopoly of Boeing, Lockheed-Martin, and other government contractor space companies will likely spur competition, decrease costs, and generate far more combinations of space power elements (and hence access) available to the nation's space efforts. However, organizational developments can also affect the military and political space power sectors as well. Foremost of these Path 5 discussions may be the military debate over an independent military space service, which we will discuss in detail later.

Like Path 2, Path 5 development impacts both the combinations level of the Grammar Delta and the transformers level of the Logic Delta of Space Power. The importance of Path 5 development lies in the culture of the new organizations that emerge in the industry and how these new industries impact innovation. New organizations are often culturally different that the old organizations and will be able to look at old problems in new ways, eventually coming up with new solutions. These new solutions (new ideas) will become manifest as new ways to combine space power elements to increase access (the combinations level of the Grammar Delta) and new ways to use the entity's raw space power (ability) to achieve economic, military, and political ends (the transformer level of the Logic Delta). Thus, new ideas generated by new organizational cultures can affect both the building blocks of space power as well as how space power can be used to achieve national goals.

Path 5 development can be stimulated by focusing on the cultures of business, military, and civil organizations involved in space power development. Ideally, as many subcultures should be nurtured as possible to allow the largest possible variety in the marketplace of ideas while simultaneously making the overall effort as economical as possible. This involves many seemingly contradictory positions. On the economic front, stimulating Path 5 development may involve promoting as much competition in the space marketplaces as possible while also allowing "natural monopolies" (i.e., monopolies achieved through economies of scale) to continue unmolested by government. On the military front, this will probably mean encouraging the creation of an independent space service (or perhaps services) while protecting terrestrial services' (Army, Navy, Marine Corps, and Air Force) space cadres from being absorbed into the new space service. The key to Path 5 development is to ensure that as many voices in space idea-making are heard as economically possible. Because it is so essential to the near-term development of space power, Path 5 development will be discussed further in Chapter 2.

Dual Concepts of Power in the General Theory

Most strategic writing considers power only in relative terms. A "standard definition" by Everett Dolman holds "that power is the capacity that A has to get B to do something B would rather not do, or to continue doing something B world prefer to stop doing, or to not begin doing something B would prefer to start."[31] This definition requires that both *A* and *B* have the capacity for deliberate decision making and purposeful action. In effect, this type of power needs two players. This relational definition of power, in the General Theory, is regarded as the logic definition of power—power concerned with the Logic Delta of Space Power. Economic, political, and military power derived from space is primarily concerned with the capacity to yield the type of power Dolman describes—the ability of *A* to get *B* to act in a way *B* would rather not. This type of power is real and important, but it is not the only type of power that exists.

The logic definition of power may appear complete when discussing relatively mature sources of power, such as the land, sea or air, where access and ability on an environmental level is virtually assured. In the land, sea, and air environments, access and ability is virtually unlimited in purely physical terms. Our technology can reach almost any spot in these mediums unless defended by other technological systems. However, space power is significantly under-developed in relation to these other mediums. Current access and ability limitations prevent us from doing many things in space without the need of purposeful denial from another conscious agent. Since technological limitations hamper space activity, a different kind of power needs to be identified to account for this weakness.

For example, consider the concept of planetary defense. Many experts believe that an asteroid or other celestial body hitting Earth caused (at least in part) the extinction of the dinosaurs and that future strikes may be large enough to destroy a major city or perhaps all life on the planet. The ability to deflect or otherwise prevent an asteroid from striking Earth would potentially be one of the most important and critical abilities humanity could ever develop, but according to Dolman's definition it could not be called "space power" because the asteroid collision wasn't deliberate. Nonsense! Of course a planetary defense capability should be considered power. Access and ability are measures of power over the environment.

Space power must also be considered in terms of the level of command over the space environment itself. When new access is available and ability is expanded, space power is increased. Remember the General Theory's definition of space power, the ability to do anything in space. Command over the environment of space, then, is considered the grammar definition of power. Thus, the grammar definition of power is the capacity for A to do anything in environment B without technological hindrance. When regarding space power, the General Theory acknowledges both the logic and grammar definitions of power.

These two seemingly different definitions of power are connected because both require the *capacity to act*.[32] While power can be relational between two deliberate actors, power can also be relational between an actor and the demands of the physical environment. A can choose to act against B and B can choose to resist or comply (the logic of power), but the capacity for *A* to reclaim resource B from the environment if *A* chooses to do so is also (grammar) power, regardless if no other deliberative body chooses to resist *A*'s action.

Role of Science in Space Power

Readers may find the lack of science or knowledge as a spire of the Logic Delta to be a major oversight of the General Theory. How can science and exploration, which are such major drivers of the current American space program, not be a part of the Logic of Space Power? The General Theory indeed accounts for the role of science in space power, but knowledge (in the form of science and exploration) is not an end unto itself as a space power goal.

In the General Theory, science and knowledge gained from exploration is one of the foundations (or principal conditions) necessary for the construction of space power. Therefore, it is a piece of the very bottom of the Grammar Delta. As a building block of space power (and only one of many), science serves only to increase the options available to designers when developing new types of space power elements: production, shipping, or colonies. While very important as a limiting factor of the quality of space power elements available to a space power (i.e., an understanding of nuclear physics allows a nation to build nuclear rockets, which are significantly more powerful than conventional chemical rockets), science is not a legitimate end to space power nor can a space program dedicated to sterile knowledge develop true space power in any significant form. Knowledge must simply be used for some purpose for it to matter.

Evidence for the inherent weakness of exploration- and science-based space programs constitutes the majority of American space history. As of 2012, no governmental human spaceflight program exits in the United States, and no significant political backing exists to re-create one. Since the end of the Apollo program, NASA has focused almost exclusively on "science" missions, with little to no regard for tangible results or return on investment to the United States beyond false color Hubble images (those posters adorning college astronomy majors' dorms bear no resemblance to their true color undoctored images) and politically unsustainable future missions to Mars. "Space exploration" and "science" are simply not able to sustain popular interest for a space program. We will discuss the pitfalls of science-centric space power later in Chapter 2.

Planetary scientist John S. Lewis describes the sterility of basic research as a goal for space programs in his book *Mining the Sky*. Lewis writes:

> I find it quite incredible that any nation on Earth would choose to devote substantial resources to [pure scientific research on the solar system] for its intellectual value alone. As a rule, governments are not intellectually inclined. If we are to return, for example, to the Moon, it will be because there is some visible relationship between that endeavor and the future material well-being of our nation and the planet. Basic research will be tolerated only if it constitutes a *balanced part of a research program that also satisfies visible economic needs*.[33]

Lewis's assertion is perfectly consistent with the General Theory. Basic research can only be a piece of a space program that relies on increasing an agent's space power—be it economic, military, or political applications (in this case, Dr. Lewis is being perhaps too narrow, as basic research may also be sustained by political and military advantage as well as economic). Basic research can only help to build sustainable space power; it is not a goal of space power in itself. Lewis continues:

> Some argue that the government must support basic research because it is the basis of the future; that all new applied science of a decade hence and all the new engineering developments of twenty years hence will build on today's basic science; that American basic science leads the world and must not be allowed to falter. These points are true [but] the scientists whose ox was gored are so "pure" that they don't know—or honestly don't care—about applied science and the commercialization process.[34]

Lewis concludes that we must "assume a future policy in which a judicious balance between long-term basic research, short-term applied research, engineering development of products, and commercialization of new products [is] reached."[35] Converted to General Theory language, Lewis's ideal future policy will encourage advances in basic research (a foundational piece of the Grammar Delta) but be committed to research, development, and fielding of new space power elements (building new elements of the Grammar Delta), thus enhancing the nation's economic space power (applied space power in the Logic Delta). Lewis again ignores political and military space power, but in all his policy recommendation fits nicely into the General Theory context.

Lewis's description of the path from basic science to a commercial product is the path from the bottom of the Grammar Delta to the end of the Logic Delta. Possony and Pournelle explain this path in detail in *The Strategy of Technology* (a monumentally important book to which we will continually refer). In their four-stage model of the technological process, an advance in basic science is only the culmination of the first phase, the "Intellectual Breakthrough." Often society must wait—sometimes over 100 years and usually "two generations"—before the science is accepted and understood in order for the breakthrough to advance even into the second stage where a practical invention using that science can be conceived and developed, much less produced and employed.[36] The technological process will be explored in detail in a later chapter. However, even this initial review makes it clear that basic research by itself cannot create space power and cannot serve as an end in itself.

Exploration is similarly an incomplete method of producing applied power, but is rather only a first step. In his landmark work on the exploration of the American West, *Exploration and Empire*, historian William H. Goetzmann writes, "In 1800 the United States was an underdeveloped land with a wilderness spread out before it, its destiny as a part of the Union still uncertain. It was the explorers, as much as anyone, who helped first secure it from international rivals, then to open it up for settlement, to lay out the lines of primary migration, locate its abundant resources, and then inquire into and point up the complex problems involved in the administration of one of the largest inland empires in history."[37] While Goetzmann lauds the role of the explorers in conquering the West, simply exploring was not enough. The land needed to be secured from rival powers (including natives), as well as be mapped for resources and migration, in order to be settled and exploited. Exploration was the first step that allowed exploitation for power purposes to commence. Had it been simply explored with no other follow-on activity (as is the case with the Moon, for instance) the West would have added nothing to the United States or its national power. Exploration, like basic science, is valuable only insofar as it is able to be exploited to contribute to national power. It is this fundamental reason that causes the General Theory to make basic science and exploration the province of the Grammar Delta, and not the Logic Delta, of Space Power.

The Logic of Science Fiction and Space Enthusiasm

When dealing with space power, it is critical for policy makers and strategists to confront the visionaries. Space power visionaries come in two different forms: the science fiction writer and the space enthusiast. Each can provide needed guidance and inspiration to the space professional and assist in the development of space power.

Why care about science fiction? Who cares about the scribblings of geek writers writing to other geeks about an impossible fantasy set hundreds of years in the future? The short answer is that we should care because science fiction is the largest repository of thought regarding the future. Before our technology takes us to places, the human imagination divines many different possible scenarios and thinks them through. The term "thought experiment" can be used to describe the mental effort to understand and categorize every problem before we begin to explore the uncharted territory beyond. The problem of how humans should best organize their efforts to conquer space is indeed a very complex problem. Great technological minds (often scientists and engineers) have considered space travel for thousands of hours over the last hundred years in agencies such as NASA, the military services, or aerospace companies. Indeed, the official compendium of space understanding has produced many wonderful technical advances.

However, technicians are very narrowly trained and understandably consider mostly technical situations when regarding space issues. Few papers on space program organization exist along the much larger volume of technical papers. Also, the number of aerospace engineers and scientists working for NASA and other organizations in the space industry number perhaps in the tens of thousands. Readership or viewership of official materials does not add a great multiplier for exposure. The "official" professionals of aerospace are few in number and write comparatively little about space organization that reach very few people.

Now consider *Star Trek*. The expanded universe of *Star Trek* that includes canon and noncanon (i.e., the "official" universe and the "unofficial" spin-offs such as books, comic books, video and board games and the like) corporately encompasses tens of thousands of hours of film and millions of pages of written material produced by thousands of individuals of various backgrounds (actors, writers, film producers, English majors, scientists, engineers, humanitarians, philosophers) that appeal to millions of people across every walk of life in almost every culture on the planet! While the latest White House or NASA policy report may be read by a few thousand policy wonks at best, the latest *Star Trek* movie was seen by tens of millions of people worldwide. There is little doubt that in worldwide contribution and exposure, *Star Trek* eclipses professional aerospace in breadth and depth of discussion on space issues. And *Star Trek* is only one example of an extremely large and detailed science fiction pantheon.

Of course, a NASA technical study of a human Mars mission will be vastly more important to technical policy than a copy of the *Star Trek Starfleet Technical Manual*, but this fact only applies to technical discussions. Many space issues are not technical issues (like the issue of space program organization, the focus of this book) and NASA engineers may be far less qualified to write on it than a political science major or historian—or a science fiction fan. It is on these "fuzzy" humanities-centric questions where the art of science fiction can be brought to bear far more effectively than technocratic NASA reports. While warp drive and inertial dampeners are not yet in humanity's technical arsenal, humanity is *already* confronted with the promise and danger of space travel. Just because *Star Trek* is based on some technology not yet achieved does not invalidate its thoughts on subjects not dependent on technology, such as how we should approach certain human conditions we may face in space. Science fiction is not just great entertainment. It is also a compendium of the thoughts of some of the world's smartest and creative minds in almost every conceivable field of labor (both technical and humanitarian) on space travel.

The most important question regarding technology is how someone will use a tool once developed, and that is not a technical question at all. Therefore, the most important question on space travel (a tool) is how humanity will use it. To answer it as best as possible, many people must approach the question from every angle imaginable. A report by a closed system such as a panel of experts will never approach the breadth and perhaps even depth of an open system where anyone interested can expand or challenge the report.

Simply put, *Star Trek* is the largest thought experiment on the future of humankind in space in world history, lasting almost half a century, spanning multiple generations, involving millions of people as producers or consumers and critics, with an almost fanatical inherent disciplined consistency. It is the clearest, most internally consistent, and rigorously explored scenario of the human future in space ever devised, and likely ever to be devised. Almost every moral, physical, or existential dilemma that has been imagined as confronting humanity from space has in some way been addressed by science fiction probably in far greater detail than any similar academic treatise if any such documents exist at all. First contact with aliens? Check. Interstellar economics and diplomacy? Check. Biology, physics, philosophy, government, economics, ethics ... almost every field of human endeavor has been considered in space travel by *Star Trek*. Science fiction at large offers, in quality and quantity, immeasurably more.

Science fiction is a vast body of thought that can be tapped to provide insight into our most pressing space question—how we will use our space tools? No one can seriously maintain that a NASA, United States, or even United Nations study on the future of humanity in space can ever be as rigorously imagined, tested, and developed as the worlds developed by science fiction.

In the General Theory, science fiction regarding space travel is valuable for two main uses. Firstly, science fiction can act as an idea generator for both the Logic and Grammar Deltas. New transformers may be able to be gleaned from speculative space fiction where writers explore how new technologies may be used in the future. New types of space power elements such as advanced propulsion systems can be inspired by an inventor seeking how to emulate his favorite prime time television show. Numerous reports of scientists and engineers attributing their professional work choice to their favorite science fiction programs attest that science fiction can stimulate space power fact. Not only can science fiction add insight into technical advancements, but different types of organizations of space institutions can be explored through the written and transmitted word in depths and detail that would be impossible to formal academic studies. In short, science fiction can inspire real results.

Secondly, science fiction can act as a "gauge of maturity" of our space efforts. If science fiction can push the boundary regarding the maturity of space power logic and grammar, it can also let us know when a nation's space power is not doing everything it can do, a check of progress. Though military space power developed through unmanned satellites has revolutionized terrestrial conflict, the science fiction ideas of populated moon bases and space stations let us know that new space power concepts and platforms *can* be developed—we are not done with building a truly *mature* space power. Science fiction, by moving the logic and grammar "bars" ever farther, shows space power practitioners both how much longer they can reach as well as how far they are away from the frontier. Therefore, science fiction acts as part inspiration for and part check on space power development.

Space enthusiasts share much in common with science fiction writers in their utility to the development of space power. Space enthusiasts are members of societies such as the Space Frontier Foundation, the Mars Society, and the better known National Space Society. Like science fiction writers, space enthusiasts continually push the limit of space power thought but also add urgency to space power projects by being perpetually dissatisfied with a program's progress. They, too, then, act as both generators of ideas as well as critics of progress. However, the main difference between science fiction writers and space enthusiasts (there is considerable overlap) is that space enthusiasts tend to be more tempered in their understanding of the state of the art in space power and can bridge the gap between science fiction fancy and the hard engineering of developing space power in the real world.

Both science fiction and space enthusiasts offer critical advantages to the developers of space power, assisting in the advancement of both space power logic and grammar. Many space power leaders should probably be readers of science fiction or space enthusiasts, or both, in order to have robust and effective space power development institutions. The General Theory recognizes the contributions of both groups to space power development, and serious professionals and space development programs ignore science fiction and space enthusiasts at the program's peril.

The Sub-Grammars of Space Power

Even though the General Theory of Space Power models the Logic and Grammar of Space Power separately, this separation is largely artificial in nature. In reality, any space project or piece of technology has both a logic and grammar component. For instance, an engine is usually not designed and built without an application in mind. Likewise, organizations commissioned to develop a certain application of space power (such as a military space force) would have a very different view of space power elements than a space-based business venture. When viewed through the lens of one of the three manifestations of applied space power (economic, political, military), the Grammar of Space Power adjusts itself slightly. The General Theory addresses this interaction by identifying three sub-grammars of space power: the grammar of war, politics, and commerce. The sub-grammar concept allows the Grammar of Space Power to be seen from a particular point of the Logic Delta. These sub-grammars are modeled as a profile view of the Grammar of Space Power with the following modifications: At the top of the delta a line extends vertically to a point above the delta. The line and point are the logic components of the applied space power. The point is the applied power (economic, political, military) and the line is its associated transformer (business plan, soft power concept, or military doctrine). The elements are also subtly shifted to address what each element (production, shipping, colonies) generally "looks like" when it is generating an applied power. Each sub-grammar has its own modified visualization as the Grammar of Commerce, Grammar of Politics, and Grammar of War (Table 1.1).

The goal of the sub-grammar Grammar of War is upholding the nation's strategic access through force of arms.[38] In our model, the grammar of war consists of the classic production, shipping, and colonies but concentrated into their more martial manifestations: treasure, bases, and naval vessels. Treas-

Grammar of Sea/Space Power: Mahan (X&Y)	Grammar of Commerce	Grammar of Politics	Grammar of War
Production (Commerce)	Trade Goods	Lines of communication	Treasure
Colonies (Bases)	Markets	Population Centers	Bases
Shipping (ships)	Merchant Marine	Treaties & Agreements	Naval Vessels (fleet)

Sea/Space Power Operational Grammar: Many Definitions, Same Element

Table 1.1 Operational Grammar of Space Power

ure is the funding and national wealth available for the military to purchase matériel and conduct combat operations (essentially the budget). Commerce generates wealth, but only the wealth channeled into military uses belong to the Grammar of War. Treasure is the lifeblood of operations, and just as commerce is the pivotal piece of the Logic and Grammar of Space Power, so is treasure the most important element of the Grammar of War. Bases take the place of colonies in war grammar. Markets and colonies serve to dissipate military effectiveness by multiplying the number of areas that must be defended. However, military bases are essential items to advance military access, provide defenses for deployed forces, and permit distant operations. Military access is determined through its bases. The final element of the Grammar of War is naval vessels (or fighting space platforms): the fighting fleet and its support ships. Naval vessels are the sword, shield, and reach of the nation's ability to project power by force of arms. They defend the strategic access of friendly commerce and prevent an adversary from access to the same. The Grammar of War dictates the strength and use of military forces to advance space power. However, being military, the Grammar of War is the least important sub-grammar of space power.

The second sub-grammar in space power is the Grammar of Politics. The goal of this grammar is to extend strategic access through diplomatic and other nonviolent means. The Grammar of Politics is comprised of political lines of communication, population centers, and treaties and agreements. Lines of communication are the currency of politics and diplomacy. As robust commerce allows the search for wealth, so do extensive lines of communication increase the probability that access will be improved through useful diplomatic ties and agreements with other entities. Population is the most important piece of colonies to the Grammar of Politics. Population centers advance diplomatic and political powers by expanding the constituency of the nation and their credibility, especially among democratic nations. Population centers not only increase political power in general, but also serve as conduits from which to exercise political power. For instance, a city near a contested area can add legitimacy to sovereignty claims against the area as well as be a center of gravity for regional politics. Finally, political grammar's "action elements," solidifying guarantees to access during peacetime and normal political interaction, are treaties and agreements with other nations. Through treaties and agreements developed through political action, national access to strategic space owned by friendly or neutral powers is assured. Treaties represent a valuable way to expand access and advance a peaceful strategic offensive (concepts described below). The Grammar of Politics is the second most important sub-grammar of space power.

The last, and most important, sub-grammar of space power is the Grammar of Commerce. The Grammar of Commerce is the mechanism with which the commerce of the Logic of Space Power is conducted. The first, and most important, element is trade goods. Without products to trade or commodities produced from the natural resources of space, no wealth can be acquired. Trade goods are the wealth of space. Trade goods can be physical products, information, or scientific knowledge. However, all wealth much enrich its owner. Material wealth is most important, and all other types of knowledge wealth must directly contribute to amassing more material wealth for it to be valuable in and of itself. The second element is markets. Markets are necessary for goods to be traded, demand to be existent, and economic activity to be conducted. The Grammar of Commerce, when viewing colonies, sees markets. The shipping component is the merchant marine or merchant astronautic corps, the fleet of ships that transport trade goods from the source of production to its market. As commerce is the most important element of both the Logic and Grammar of Space Power, the sub-grammar of commerce is the most important of the sub-grammars.

The sub-grammars are the building blocks necessary to develop the Grammar of Space Power in support of the Logic of Space Power. For true space power to be developed, all of the elements of these sub-grammars must be developed in proportion to their need to support the overall space effort. In all things, the goal of space power is to generate wealth from space. The proper proportion of the elements of space power is that which maximizes the wealth that can be generated from space in both peace and war. None of the elements of the sub-grammars should be placed in higher importance than the Logic of Space Power. The sub-grammars are all composed of the traditional space power elements: production, colonies, and shipping. Knowing that each grammar places a particular spin on the space power elements helps us to better understand the fundamental necessity of any sea or space power—*access*.

The sub-grammars of space power can be used to advantage in two specific applications. The first is to conduct a holistic analysis of a specific program to advance a particular application of space power such as a military space service. A military space service would undoubtedly view space power grammar with an eye for building military, as opposed to commercial, equipment. The second application is to analyze a "heresy of space power"—when a sub-grammar inserts itself as the totality of the Logic of Space Power. The space power heresies are described later.

Generally, an application-specific organization would not view space power purely through the Grammar and Logic of Space Power. They would view grammar through their particular logic, namely their ends (applied space power) and ways (their unique transformers). This unique grammar would tend to change the particulars of production, shipping, and colonies. A military space service, as an example, would tend to see treasure (funding), warships, and bases rather than the generic elements of space power.

A particularly valuable use of the sub-grammar matrix is to take a complete system built for a specific application (for instance, a commercial space system) and place its component elements in a graphical representation of its original sub-grammar. Then, adjust that sub-grammar using a different sub-grammar and see if this new construct provides a valuable new type of access. For instance, a remote-controlled International Space Station cargo vessel is developed for a specific commercial application (space station replenishment transportation). If we then apply the Grammar of War to the system, we can examine what it can do using military doctrine and concepts. Can this transport be used for orbital reconnaissance, for instance? Essentially, this analysis technique can be used to assess military applications of commercial equipment, and vice versa, a promotion of Path 2 (spillover) development in the General Theory.

Heresies of Space Power

Most, if not all, problems encountered in space policy can be attributed to substituting the Grammar of Space Power (commerce, politics, or war subgrammars) for the Logic of Space Power. Much as military officers confuse tactics for strategy, space policymakers often elevate their parochial concerns of grammar above that of the Logic of Space Power, thereby frustrating their actions and the goals of the nation at large. People can elevate any of the grammars, and each of the commonly acknowledged missteps in the American space effort can easily be explained as a "heresy of space power" that elevated space power grammar above the logic. Indeed, rarely has the Logic of Space Power been identified with any of the major efforts in American space power. Most of the major activities in the civil and military space sector has been examples of good grammar and bad logic.

The Grammar of Politics has been the grammar most often substituted for the Logic of Space Power in American policy, and it continues to exercise undue influence in space thinking. The political heresy (political grammar substituted for logic) focuses on building a political element of space power as a goal unto itself. This can take the form of a treaty, an agreement, a project, or even an idea with a goal that is detrimental or indifferent to generating wealth from space. Examples of the political heresy are numerous and many space constituents fall into its trap. The construction of the International Space Station is such a heresy. Instead of building a space station to expand access to space, the ISS instead became a project with the primary goal of "international cooperation" and was showcased as a diplomatic triumph (cynics called it a State Department welfare program) instead of an economic disaster. Indeed, the entire concept that space exploration and settlement is "too expensive" for anything other than a multinational effort is unproven at best and a blatant fiction at worst: an example of the political goal of multinationalism being elevated above the goal of generating wealth from space.

Other examples of the political heresy are so-called flags and footprints missions such as the Apollo program and the proposed Mars missions. Many space enthusiasts rightly call them boondoggles and dramatic wastes of resources. The Apollo program did not advance a spacefaring society because it was designed to beat the Soviet Union in a high-stakes popularity contest rather than expand strategic access to space. Political goals such as "prestige" and "soft power" are transitory political currency and are rarely adequate returns for the resources invested to gain them. Treaties such as the Outer Space Treaty stating that space resources are the "common heritage of mankind" are disastrous to generating wealth from space and inhibit space commerce almost entirely. Nonetheless, misguided space advocates often tout these monstrosities as great victories, dooming mankind to eternal bondage to their home planet by stifling the ability to build sustainable space projects that adhere to the Logic of Space Power.

Political heresies are not the only missteps that can be taken in space power policy. The second great heresy is to replace space power logic with the grammar of war: the favored error of the war hawks. The war heresy tends to be that of viewing space power in absolutely fighting terms and to resist space power growth for the risk it has in upsetting the current military balance of power. The paranoia in the late 1990s and early 2000s over so-called dual use space systems in military and defense circles and subsequent attempts to restrict wide classes of technological exports "vital to national security" through legislation are prime examples of this error. There is no doubt that restrictive export controls severely damage the American space industry, and many blame them for the precipitous collapse of standing in the international world of American space companies in the last decade. The war heresy sees American preeminence in military space as something that needs to be protected by attempting to freeze development to the "status quo" and refuses to see the advantages of freeing commerce to build wealth from space. Often, this obsession with the status quo so diminishes the military's understanding of space power that generals can speak of having achieved "operational mastery" of space already with a straight face, when almost any enthusiast knows that human access to space is almost laughably limited. Following the Logic of Space Power is deemed too dangerous and American space power is stalled in development due to timid national security leaders that misread the fundamental advantage of being a space power nation. This is the sadness of the war heresy.

Even though commerce has preeminence in both the Logic and Grammar of Space Power, slavish devotion to the grammar of commerce also leads to space power heresy. The commerce heresy appears when devotees of space commerce ignore or deny the proper role of balanced politics and warfare in space power. This heresy has been the least influential to space policy but has an unhealthy influence on the space enthusiast community. The commerce heresy is most observant in the cries against a military presence in space (characterized as "militarization" or "weaponization") because it may discourage investment in space. This is a position taken by some space advocacy groups, and the war heresy could indeed endanger future space activities by making space a war zone without purpose, but the military does have a proper role in space. According to the Logic of Space Power, the military exists to defend strategic access to the space environment and deny it to adversaries in time of war. Both an offensive and a defensive role are legitimate. However, as the military must be balanced against being too aggressive, it must also avoid being too timid.

The commerce heresy also extends to political matters. Often, free-market space advocates deny the legitimacy of taxation, regulation, or legal jurisdiction over space activities. They have a point to an extent, but government space support such as emergency services or traffic monitoring must be paid for through taxes, citizens' safety must be ensured as much as practical and efficient through prudent regulation, and basic rights and legal protection must be extended into the space realm. Leviathan can destroy space power through bureaucratization and stifling creativity and innovation, but in its proper role political effort can be a valuable tool in expanding the Logic of Space Power.

The Logic of Space Power can only be advanced by its constituent parts: economic, political, and military space power, in their proper role. The heresies created by substituting grammar for logic have held back American space power from its inception. Many of the disappointments can be described as a failure of policy makers to respect the Logic of Space Power. Using the constructs of the Logic and Grammar of Space Power, we can easily examine policies and projects to determine if they will help build wealth from space and advance the cause of American space power.

Applying the Space Power Deltas

But the Logic and Grammar Deltas can do far more than help identify fallacious policies that inhibit space power. Nations and other wielders of space power can also use the deltas to design enlightened policies that will enhance the effectiveness of their space programs. Because they model significantly different aspects of space power development, both the Logic and Grammar Deltas suggest unique activities make space programs more efficient. We will discuss how to optimize Grammar Delta activities in Chapter 3, but first we will discuss methods with which we can optimize operations along the Logic Delta of Space Power.

Chapter 2

Organizing for Effective Development—Logic

With the theory of space development described, this chapter focuses on the Logic Delta and what can be done to strengthen a nation's ability to develop space power through effective space power logic. The key to space power logic is to perceive the criticality of organization to the success of a nation's space program. We begin by examining the importance of organization theoretically, then explore the current American space program, and consider an intriguing case study that showcases the role of organization to space power development.

The Importance of Organization

Development is created through new combinations of space power elements and ways to apply them, through using the five paths. Critical to the combination of the elements of space power are the natures of the institutions that manage the elements. Therefore, the General Theory treats the organization of space power institutions as an important factor in the development of space power. An example of Schumpeter's fifth path to development is, indeed, simply a new organization that will proceed with exploring new combinations!

Schumpeter's mechanism of change for every path is the entrepreneur. Entrepreneurs are defined as anyone in pursuit of new enterprises, which themselves simply are the carrying out of new combinations.[1] Entrepreneurs do not need to be independent businessmen such as venture capitalists, as in the common understanding of the term. Entrepreneurs are people who carry out new combinations regardless of their personal station.[2] They can be business owners or employees and from the private or public sector. They can be scientist, engineer, businessman, politician, statesman, military strategist or tactician. And in the General Theory of Space Power, the entrepreneur works on either the transformer (logic) or combination (grammar) level. The entrepreneur is characterized by initiative, authority, and foresight.[3] The entrepreneur is the innovator who allows any type of development to happen.

Innovation is championed in the private sector. Business and technology leaders have elevated innovation to heights that sometimes dangerously approach a panacea. However, while innovation is often trumpeted in senior leadership speeches, actually encouraging and harnessing innovation in the military remains an intractable problem. The military remains a very conservative and hierarchical organization that finds little use for military strategic or tactical entrepreneurs, aside from occasional declarations of officers for posthumous sainthood for being "visionary" or before their time (fully realizing that "before their time" simply means before the military was finally forced to agree they were right!).

In order to be innovative and allow development in its field to occur, the military must find a place for the entrepreneur in uniform. Military entrepreneurs have existed in history and some have helped their nation survive and thrive through trying periods. Even if resisted, military entrepreneurs are vitally important. However, while the corporate military does little to encourage entrepreneurship, individual officers have often become the military entrepreneur's greatest champions.

Major General I.B. Holley, Duke historian and Air Force Reserve officer, spent his life exploring the nexus of technological and doctrinal development in military adaptation and his own thoughts are in ready agreement with Schumpeterian development language and the General Theory of Space Power. Holley identifies one of the most important concerns of doctrine development in the military:

> How best can we ensure that suitable doctrine is developed for radically new hardware, novel weapons, made possible by the application of hitherto unexploited technology [Path 2 development]? Here the path is strewn with obstacles. We design tests and conduct maneuvers to try out the new weapon; given our strong human propensity to lean on previous experience, how can we avoid designing a test that reflects our past experience rather than seeking the full potential of the innovation? When the results of our tests and maneuvers are recorded, how can we ensure that preconceptions and prejudice or partisan branch or service interests do not distort the substance of our reports? Can we be sure that institutional bias isn't coloring our findings?[4]

It is clear that Holley's concerns are intimately intertwined with military development. Holley believes that these concerns can be best addressed by building a "truly effective organization for formulating doctrine ... that is staffed with the best possible personnel."[5] Holley continues:

> What is a sound organization? Ultimately, no organization is better than the procedures designed to make it function. Yet, on every hand in the armed forces today, we see men in authority assigning missions and appointing leaders to fill boxes on the wiring diagram while seriously scanting the always vital matter of internal procedures. It is the traditional role of command to tell subordinates *what* to do but *not how* to do it; nonetheless, it is still the obligation of those in authority to ensure that the internal procedure devised by their subordinates meets the test of adequacy.
>
> And what do we mean by the best people? We must have officers who habitually and routinely insist on objectivity in their own thinking and in that of their subordinates. This does not rule out imagination and speculation by any means. But we must have officers who insist on hard evidence based on experience or experiment in support of every inference they draw and every conclusion they reach.
>
> We need officers who will go out of their way to seek and welcome evidence that seems to confute or contradict the received wisdom of their own most cherished beliefs. In short, we need officers who understand that the brash and barely respectful subordinate who is forever making waves [the entrepreneur] by challenging the prevailing posture may prove the most valuable man in the organization—if he is listened to and providing his imagination and creativity can be disciplined by the mandate that he present his views dispassionately and objectively.[6]

Of course, even with a sound organization and the best people developing doctrine (the military transformer), they will be worthless unless the parent organization will give their ideas a fair hearing. This is not always the case as historically effective doctrine organizations (such as the Air Corps Tactical School) have had considerable difficulty in "institutionalizing" their developments into the combat forces and parent institutions at large. Even with solid innovations being developed, they must still be reconciled with tradition and integrated into the organization they were meant to improve.

A key insight to reconciling innovation and tradition in the military can be gained from the Original Institutionalist School of economics, founded by thinkers such as Thorstein Veblen. Institutional economists identify two classes of values in a culture—instrumental values and ceremonial values. Instrumental values concern solving the problem an institutional culture (such as the military culture) was formed to solve. They tend to favor the application of new tools, knowledge, and skills into the institution's problem solving process. Ceremonial values, alternatively, are values that form the tradition and social structure of the institution. Ceremonial values are generally ambivalent to problem solving but generally oppose incorporating new technologies "that could threaten existing social relations with respect to power, wealth, position, etc." in the institution.[7]

The application of innovation (whether in technologies or doctrine) in organizations is dictated by the interplay between the institution's instrumental and ceremonial values as they struggle to reconcile and incorporate the new knowledge the innovation represents. "If the ceremonial values are eventually pushed aside in favor of instrumental values, institutions adapt to the new circumstances and progress is achieved. If the ceremonial values continue to shape behavior, ceremonial values trump instrumental values and the institutional patterns (justified by the ceremonial folkways) are said to be ceremonially encapsulated."[8] The dilemma of instrumental versus ceremonial values are well known in the military. As strategist J.F.C. Fuller said, "To establish a new invention is like establishing a new religion—it usually demands the conversion or destruction of an entire priesthood."[9]

Institutional economists tend to imply that instrumental values are "good," leading to innovation and development, while ceremonial values are "bad," entrenching illegitimate power and creating "invidious distinctions" in society. While traditions and entrenched business interests may not be ideal in the business world, even the most strident military innovators (the Air Corps Tactical School's motto was "We make progress unhindered by custom") must acknowledge that custom and tradition in some form has legitimate value. But not all.

Few military professionals would throw out all military tradition for innovative efficiency, and it would be a disaster if anyone tried. The partitioning of personnel between officer and enlisted ranks is probably a positive and permanent innovation in military development (though how officers are traditionally selected may not be as sacrosanct!), but dismissing a technological innovation such as longbows because the traditional ruling class (such as knights) would be rendered obsolete is far harder to sanction. Military organizational planners must strike a fair and enlightened balance between those ceremonial values that guard the soul and embolden the spirit of the institution from those that simply favor a temporary and disposable mini-elite.

The distinction may be in the eye of the beholder, and we can better understand ceremonial values in the military as those values that defend Holley's procedural "test of adequacy." In the military, innovation should be guilty until proven innocent, and to prove an innovation an improvement is the challenge that good ceremonial values should present to instrumental challenge. The healthy interplay of instrumental and ceremonial values in a good organization may best be described by General Sir John Burnett-Stuart's words to B.H. Liddell Hart shortly after becoming commander of the experimental

British armored force in 1926: "It's no use just handing over to an ordinary Division commander like myself. *You must [assign] ... as many experts and visionaries as you can; it doesn't matter how wild their views are if only they have a touch of divine fire. I will supply the common sense of advanced middle age.*"[10]

Vision is the blueprint for developing the future. Vision is the upper limit of space development's potential. Space hardware and physical infrastructure is extremely expensive while dreaming is virtually free. However, choosing a proper vision for what future space power should achieve and how it can achieve it is far more important because it is what channels all other material support into valuable (or wasteful) action. Therefore, the critical and central component of a successful organization is its vision for the future.

Three (or Four) Visions of Space Development

Space advocates have not always agreed on what the goals of the space program should be, and what the human future in space should look like. In short, not all space visions share the same dream. The differences are important, and sometimes visions differ more than what they have in common. By discussing the various major space development visions that have been proffered throughout the decades of the Space Age, we can gain a better foundation for discussing the future of space power and how to realistically begin to build a vision of mature space power.

This book argues that three major visions of space development have been offered, all are generally attributable to major thinkers in the space community, and all are broadly represented by advocacy organizations dedicated to furthering their particular vision. Of note, two have had considerable sway over the American space program, while the other has captured the imagination of the majority of grass-roots space advocacy organizations. I will call these three the von Braunian vision (named for the famous rocketeer Dr. Wernher von Braun), the Saganite vision (named for legendary scientist Dr. Carl Sagan) and the third, the O'Neillian vision (named for the lesser-known, but highly celebrated in the space field, space station design pioneer Dr. Gerard K. O'Neill). Into this mix of competing ideas I will add a fourth vision that made a great impact in the Cold War and was championed by many highly acclaimed space thinkers, but has been unduly forgotten by history. This vision is the Grahamian vision, named for Lieutenant General Daniel O. Graham, its primary architect. Each of these visions presents an inspiring and positive future of humanity in space, but each emphasizes different points of concentration and offers different overarching goals. By understanding each of them, we can begin to use them as departure points and a common language with which to discuss building a future in space with fully developed and powerful space power at the end of the journey.

FLAGS, FOOTPRINTS AND TECHNOLOGICAL CONQUEST—THE VON BRAUNIAN VISION

Perhaps no one has had a more commanding presence in the space effort, or has been more controversial, than Dr. Wernher von Braun (1912–1977). The chief designer of the German V-2 rocket in World War II, von Braun created arguably the first space weapon. Twenty years later, he was the towering figure and erstwhile leader of the American space program, first at the U.S. Army's Redstone Arsenal and later NASA. In his tenure he created the first vehicle that could carry human beings to another world, the *Apollo-Saturn V* system. In between, he popularized the concept of mass human space travel to the American public in an influential series of articles in the magazine *Collier's Weekly*.

Von Braun's vision for space development is so widely accepted that many believe it is the *only* vision and is universally accepted. For our purposes, his vision is best explained in his little-known fiction book *Project Mars: A Technical Tale* (written in 1948 but not published until 2006—the consensus being that von Braun was a better engineer than novelist), or its more well-known technical appendix, which was published earlier as *The Mars Project* (1952). This book introduces the key points and tenets of the von Braunian vision, which will be familiar to those who have followed the space program for any amount of time.

In his book, von Braun describes a large, manned conquest of the planet Mars in the mid–1960s. A fleet of ten spacecraft, constructed in Earth orbit and serviced by space stations and reusable space shuttles, would carry 70 explorers to Mars for a 443-day ground mission before returning to Earth.[11] This large-scale building approach to accomplish a grand mission has been almost indelibly stamped into NASA thinking since its inception. Von Braun's scenario is still the benchmark for Mars mission planning. However, the von Braunian vision is not simply a large-scale exploration of Mars. It is through his explanation of how we will accomplish it that we find the biases, assumptions, and tenets of the von Braunian vision.

The von Braunian vision is best described as a government-led, mission-oriented approach meant to focus the space program on one overarching goal. It is in many respects a space approach that has much in common with the large Antarctic expeditions of the time (U.S. Navy Operation Highjump, in particular). The von Braunian approach to space development was to take a specific grand destination and mission—in this case a 70-person, 443-day exploration of the surface of Mars—and take a stepping-stone approach to developing the matériel necessary to accomplishing the mission. First, reusable space shuttles will be built to construct space stations. Then these space stations will build the Mars spacecraft. Then the spacecraft will go to Mars. The astronauts will explore Mars. The astronauts will return. Mission accomplished. Space program moves to different grand mission. It is a vision of technological conquest. The Mars mission is a goal in and of itself, and the whole reason for going is to prove that we have the technology to do it. Science, exploration, or exploitation takes a back seat to the sheer act of going.

Delving deeper, the von Braunian vision champions a command economy where national (or international) resources are devoted to accomplishing the grand mission. Only that which is necessary to accomplish the mission is built. The Apollo program to reach the Moon was the von Braunian vision in action. The U.S. government, in the guise of NASA, purchased rockets designed and built by government contractors to send government astronauts to explore the Moon and return. In many ways NASA still works under the von Braunian vision. The government space shuttle was operated (until 2012, with no current replacement) by government contractors to mostly service the government International Space Station (ISS) staffed by government astronauts. For the most part, the ISS exists mostly to simply exist. Proponents are correct to say that constructing the ISS has given engineers a great deal of experience in constructing large space structures, but rarely do these people tell us where we will use this new knowledge in the future. Not their department.

The von Braunian vision is often derided by critics as "flags and footprints." Mostly due to disappointment that Apollo did not herald an age of lunar colonization as well as the eventual realization that Apollo as a government program was never intended to do so, these critics point out a very major negative aspect of the von Braunian vision. In the von Braunian vision, if the government does not want or pay to do something, it doesn't get done. More importantly, any government equipment built for the government mission will have little utility to anyone not using it exactly as directed. For instance, the Apollo program did not build a bridge to the Moon, nor leave us with any usable infrastructure in orbit or on Earth (short of static displays of rockets at NASA centers) to make future expeditions easier. Apollo gave us nothing but American flags and footprints on the Moon (before the Soviets), which was all that Apollo was intended to do. It is important to remember, before we criticize Apollo too much, that von Braun's Mars mission promised the same thing. There was no permanent colonization effort envisioned, and no private use of the space facilities to make a profit. Critics should have known better than expect anything from Apollo. The von Braunian vision is clear—with the right money and government support, the (narrow) mission will get done. Everything else is an ignorable and expensive distraction.

NASA and many of its supporters still labor under the von Braunian vision. Government control of the space program is championed at NASA, and commercial efforts are generally ignored even if they are not exactly discouraged. Government centers still would prefer to build their own equipment or use favored contractors rather than commercial alternatives. The mission is everything, and anything not of the mission in superfluous. Just as the von Braunian approach was used in Antarctica, so are its fruits the same—neither Antarctica nor the Moon have enjoyed any significant development since the government completed its large exploratory expeditions. Space enthusiasts have noticed, and the pitfalls of the von Braunian vision led to the development of alternative space visions.

Science, Exploration and Pale Blue Dots— The Saganite Vision

In the early 1970s, the drawbacks of the von Braunian vision of government human mission dominance of the space program were not yet commonly understood. However, a different vision for human action in space was developing. This was to use automated, robotic probes to explore the solar system and eschew the very expensive (and unrewarding, thanks to von Braunian mission myopia) human spaceflight efforts currently in vogue. Although there were many advocates of this vision (Dr. James Van Allen being notable), Dr. Carl Sagan (1934–1996) was the foremost proponent of this vision's concepts.[12] Sagan envisioned a dual space effort to explore the solar system using robotic probes and search for extraterrestrial intelligence (SETI) through radio waves rather than starships. His ideas found wide expression in his books *Pale Blue Dot* (1994) and *Contact* (1985). His thoughts on the primacy of robotic exploration and science-driven space activity have a large following in the non-profit advocacy group the Planetary Society (which he helped found) and NASA's own Jet Propulsion Laboratory, the capital of American unmanned space explorers. Adherents to the Saganite vision are often space scientists who call human space missions prohibitively expensive (Van Allen, the space scientist who discovered the radiation belts that bear his name, was a harsh critic of human spaceflight in his later life, as too costly, unnecessary, and trivial) and prefer robotic explorers to advance science as the ultimate goal of space flight. Some have even called human spaceflight a technology that needs to be uninvented.

Also inherent in the Saganite vision is to keep space pristine from any activity other than "noble" scientific endeavors. Making money by mining an asteroid is less uplifting than sending a scientific probe to Pluto. This thought can be considered an unfortunate bastardization of Sagan's original, and much more reasonable, insistence that if Martian life-forms are found (even in microbial form) Mars should be left to them and preserved against human encroachment, a reasonable position even if it isn't widely shared.

The Saganite vision pictures humans staying safely home, leaving the heavens to robotic probes to scour the void. Probes like the *Pioneers* or the *Voyagers* should be sent to every stellar nook and cranny to learn as much as possible. Even interstellar exploration is possible through von Neumann probes (named for the computer theorist John von Neumann) small, self-replicating robots sent across the universe to explore without risking human lives. Searching for alien intelligences is left to radio telescopes and other methods of remote detection. Indeed, a fundamental tenet of the Saganite vision is that faster-than-light travel is impossible regardless of the technological advancement of any life-form or civilization, and therefore life would not extend beyond their home star system at the very most.

To be fair, Sagan himself was a proponent of expanding the human presence in space and extending the human biosphere. I unfortunately must use "Saganite" to describe the antihuman spaceflight vision because current proponents of ending the human space program in order to free money for more probes (ostensibly because science is all we can accomplish or want to do in space) have clung to Sagan as a guiding saint.

The Saganite vision does have a human spaceflight component, but it is very specific and bounded. Sagan devoted a great deal of professional thought to nuclear warfare and how to protect human civilization. He was an early and outspoken proponent of what would later be known as "planetary defense"—advocating the search for asteroids and other natural space dangers which may threaten life on Earth. He did, after all, say that the dinosaurs became extinct for lack of a space program. The Saganite believes that if humans must travel into space, they should do so primarily as a type of life insurance to protect against cataclysm (natural or artificial) on their home world.

Needless to say, the Saganite vision (especially its Van Allen–inspired extremists) takes a great deal of heat from people who believe that human spaceflight is a good thing. However, the Saganite impulse is strong among many proponents of spaceflight in that many believe science is the most important thing the space program can do. This is far less controversial, but those that disagree with science fetishists tend to be very outspoken against another myopia. Resisting the perceived governmental and technocratic myopia of von Braunians and the compulsive obsession with science of the Saganite, a different vision was born in an effort to become more balanced and more relevant to the human experience.

INDUSTRY, COLONY AND FREEDOM— THE O'NEILLIAN VISION

Also in the 1970s, the ending of the Apollo era and the burgeoning hope for expanded access to space from the proposed space shuttle drove space thinkers to offer a new vision of rapid human space development and widespread human colonization. Terrestrial pessimism over the energy crisis, oil shortages, and overpopulation and famines mixed with the optimism of space exploitation to drive new ways of solving Earth's problems with space technology. One of the foremost proponents of solving Earth's problems with space expansion was physicist Gerard K. O'Neill of Princeton's Institute for Advanced Study. His powerful vision was presented to the public in his book *The High Frontier: Human Colonies in Space* (1976). In it, he claimed that free enterprise, space industrialization, lunar and asteroid mining, and mass driver space launch systems could be used to build solar power satellites to beam the Sun's power to Earth (thus eliminating the need for oil and other fossil fuels) and build gigantic orbital space colonies to eliminate population problems on Earth. Although no governmental organization has ever embraced the O'Neillian vision (though certain elements in NASA have reacted favorably with studies and other efforts on O'Neillian programs), it has a very large following in the space advocacy organizations, specifically O'Neill's own academic Space Studies Institute (SSI) and the libertarian-leaning Space Frontier Foundation.

Essentially, the O'Neillian vision promotes turning space into a *human environment*. Not satisfied as keeping space as a playground for von Braunian bureaucrats or ivory tower Saganite scientists, the O'Neillian vision sees hundreds, then thousands, of people living, working, loving, and raising families in space—people who think of space as *their home*, not simply an interesting place to visit. There, individuals are free to chart their own destiny and work together or solo to achieve their own "space dream" (a 21st-century change of scenery version of the American dream). Since space is a human environment, we will carry our organizations and cultures with us, without one central bureaucracy managing a monolithic "space program." Most O'Neillians recognize the legitimacy of government space operations (some space libertarians dissenting), but government takes a supporting, not dominant, role.

Others have advanced books and ideas so similar as to be allied with the O'Neillian vision and add significantly to its depth. Dr. John S. Lewis of the University of Arizona wrote about the massive amount of raw materials that exist in the solar system in *Mining the Sky* (1996), and lunar industrialization has been explored in Apollo astronaut Dr. Harrison Schmitt's *Return to the Moon* (2006). The O'Neillian vision has surpassed Dr. O'Neill and become a consistent call for industrialization of space in many different ways using many different organizations. This large body of work is coming very close to a "critical mass" of ideas that can sustain a robust commercial enterprise and competition in space.

The O'Neillian vision has been particularly embraced by spaceborne libertarians such as the Space Frontier Foundation. This vision draws libertarians for the freedom and opportunity it offers. With many different "O'Neill cylinder" space stations acting as orbital cities (orbital nations?), many different experiments with social structures and communal decisions can exist in relative peace with its neighbors free of outside interference, if desired. The sophisticated space colonist can pick and choose his or her preferred lifestyle among a veritable universe of possibilities. A shrewd businessman can also make a fortune taming a frontier that is, quite literally on a human scale, infinite. However, this libertarian streak (however admirable) has often convinced O'Neillian advocates that government is the enemy of space flight. This belief may be harmful to finding a healthy balance between private and public sectors in space travel, and pose a few problems for considering how to organize the government's operations in space. Building a real Starfleet will be much more difficult if the champions of space development deny the legitimacy of government space operations entirely. This quirk of the O'Neillian vision leads us to consider a fourth space development vision, one that seeks to find a balance between public and private expertise in space development.

Public and Private Efforts for National Power—The Grahamian Vision

The 1970s space development theories were imbued with the opportunities and concerns of their decade. The Saganite vision was a reaction to the impressive scientific data returned from solar system probes and ensuring human civilization's continuance in the face of nuclear war or asteroid strike. The O'Neillian vision was founded upon the space advocate's optimism in the future of the space shuttle and confronting the problems of energy and overpopulation. In the 1980s, a new space development vision emerged that attempted to lift the United States out of the cultural malaise of the Carter years and rally to defeat the Soviet Union and end the Cold War. It was envisioned by U.S. Army intelligence officer Lieutenant General Daniel O. Graham. He insisted that the United States could press its technological advantage in space operations to secure the "high ground" of space both militarily and economically. Militarily, satellites could be placed in orbit that could defeat a sizable portion of the Soviet nuclear missile forces in space while economically, U.S. companies could use space to develop superior products such as pharmaceuticals and building materials. In the short run, his "High Frontier" strategy would force the Soviets to bankrupt themselves in a futile attempt to keep up. In the long run, renewed national interest in space would secure a robust U.S. high-tech economy and make the entire world richer through using space for human benefit and show the superiority of freedom over centralization. He outlined his vision in the book *High Frontier: A New National Security Strategy* (1984).

Though many have probably never heard of "High Frontier" or General Graham, his work has had significant impact on the U.S. space effort. His ideas on missile defense were developed (some might say perverted) into the Strategic Defense Initiative (SDI), better known as "Star Wars." His lobbying for cheap, easy space launch was the key factor in the construction and test flight of the Delta Clipper (DC-X) reusable rocket, a vehicle held in high regard by many space advocates. Allies of his theory counted many famous space personalities, such as science fiction writers Arthur C. Clarke and Jerry Pournelle. It is a very powerful vision.

The Grahamian vision at first glance seems very warlike and militaristic, since it quite frankly asserts as its goal the extension of national power to whoever uses its strategy. This shouldn't come as a surprise, considering that its genesis was as a way to decisively win the Cold War and it was written by a high-ranking military officer. However, to conceive of it as a "war hawk" strategy only is to embrace a misconception. The Grahamian vision is a "geopolitical realist" vision for space development. The realist school of strategic thought simply believes that states will generally act in ways that best enhance their power or security, rather than working primarily for ethical or idealistic considerations. The Grahamian vision is a "space power" vision, where the nation uses space to achieve its own interests—to build its economy, add to its defense, and secure the blessings of space to its people. There is nothing inherently warlike about it. Nations can coexist peacefully in exploiting space (it is certainly large enough for everyone!), though it does deal in unpleasant realities that *Star Trek*–style space utopians sometimes avoid.

In fact, aside from the Grahamian recognition that national interest is a significant force in space development, this vision is not significantly different from the O'Neillian vision. Private actors are encouraged to go forth and generate wealth from space. However, in the Grahamian vision, every couple of space stations will be a military outpost. The military need not be offensive in any manner, either. Remember, the Grahamian vision's space military presence was intended to eliminate the threat of nuclear annihilation, not to conquer the world. And, as *Star Trek* always shows but fans often need to be reminded, the most popular science fiction space visions have a very large military presence in space, and that is often positive. Therefore, perhaps the Grahamian vision is simply the O'Neillian vision modified to recognize that the military and sovereign nations will still be with us in space. (Indeed, both O'Neill and Graham named their foundational books *High Frontier*.)

A more recent proponent of national space power, perhaps best described as a neo–Grahamian, is U.S. Air Force Air University strategic studies professor Dr. Everett C. Dolman. In his book *Astropolitik: Classical Geopolitics in the Space Age* (2001), he outlines a "realist school" strategy of space development to enhance both the soft and hard power of the United States through military dominance of space and economic expansion through space, a plan which will be discussed in Chapter 5.

Emphasis is being placed on the Grahamian vision because it has been largely forgotten. Even though it was written by a military officer and had a large impact on Cold War military spending, it is important to note that the military space commands do not subscribe to this vision in any meaningful sense. The space commands are more interested in using established space technology to enhance traditional war fighting methods (like providing navigation, imagery, and communications data) rather than advance space development in any significant fashion. Although every now and then a treatise on the future in space emerges from military circles that approach space development, it is a fair assessment to say that, by and large, the military does not promote space development in the space advocate's sense and that no organization really champions the Grahamian vision.

Whether or not the Grahamian vision should be considered a space development vision in its own right or simply a "realist" perspective on the O'Neillian vision is an open question. Perhaps it doesn't really matter. What does matter is that General Graham and his take on the "High Frontier" should be remembered as the insightful, profound, and important space development idea as it is.

These four visions are presented in order to prove a point: that the General Theory of Space Power doesn't necessarily endorse any of them. However, the general theory can be used to explore each vision's potential shortcomings in order to understand where pitfalls might exist in each offering and offer suggestions to correct vision policies in order to extend each vision's goals. We will now explore the current American space program to assess its organizational effectiveness.

The American Space Effort: Stuck on Impulse

Most people, even those not interested in space affairs, sense that the American space effort isn't healthy. The last time any human visited the Moon was forty years ago. The years 2001 and 2010 have both come and gone, without anything in space even remotely close to the wheeled space stations, Moon bases, or powerful Jovian space cruisers of the movies *2001: A Space Odyssey* (1968) or *2010: The Year We Make Contact* (1984). The space shuttle that was widely hailed as a "space truck" that would open up space to large-scale development in such books as T.A. Heppenheimer's *Toward Distant Suns* and G.H. Stine's *The Third Industrial Revolution* (both 1979) instead turned out to be the most expensive and complex human machine ever devised, and probably kept space development and colonization an unrealizable goal for at least another generation. Beyond these simple generalizations, we can measure the lack of space development over the last fifty years since Yuri Gagarin's first flight with three metrics: the number of people who have visited space, the distances humans have traveled in space, and the cost of sending a pound of payload into orbit.

Only about 500 people have traveled into space since the beginning of the Space Age to 2010. The space shuttle allowed up to seven astronauts at a time and has enlarged the number considerably in the 1980s and '90s, but this number is about the same as a fully loaded Boeing 747. In fifty years, this is not a healthy sign. With regard to the distances humans have traveled into space, only 24 have gone beyond low Earth orbit (gone higher than a few hundred miles up in altitude)—all of whom went to the Moon. Those 24 to leave Earth orbit all completed their journeys by 1972. In almost forty years, no one has traveled farther into space than the distance from Los Angeles to Phoenix. The cost to orbit, a very important economic metric to heavy space development, is perhaps the most essential metric because it directly affects how expensive space projects will be. In the 1960s, rockets cost around $10,000 per pound of payload into low Earth orbit. Today, there has been very little improvement and the cost to orbit is higher today than in the beginning of the Space Age, even accounting for monetary inflation! There have been advances in space travel, to be sure. We now have fielded fleets of satellites for communication, imagery, and navigational uses. The International Space Station (ISS), though far inferior to the spinning space stations of the 1950s and '60s imagination, has provided a "permanent" human presence in space, if only for a three- to seven-person maintenance crew. These comparatively paltry successes provide small consolation to people who wish to see a future in space similar to that depicted in *Star Trek*.

It is clear using any metric that space has not been developed as fast as some would have hoped and many believe possible. Many reasons and excuses have been advanced to explain or dismiss this simple observation. In order to understand and consider the truth of these reasons, we need to look at the organizations charged with managing the American space effort. The two main government agencies that manage the American space effort (the efforts of the private space sector will be discussed later) are the National Aeronautics and Space Administration (NASA) and the Department of Defense (DOD) and its suborganizations such as the Air Force, Army, and Navy Space Commands, as well as other smaller agencies. In fact, this dual nature of the government program, with independent organizations for civil-science and military space efforts, will be found to be a possible cause of America's relative lack of advancement. However, we must briefly look at each of the agencies (we will explore them in depth in a later chapter), which we will call NASA and the Space Commands, separately in order to understand America's space story so far.

NATIONAL AERONAUTICS AND SPACE ADMINISTRATION

There is no doubt that when an American thinks of the space program, the overwhelming majority of people will immediately think of NASA. NASA conducts all of the American government's human space missions (even those with military missions attached) and operated the space shuttle and still operates the American portion of the ISS. NASA's automated probes have sent explorers beyond the farthest planets and have explored many of them in the solar system in depth. NASA was created to administer the government's civil space efforts. Most assume that part of the civil space effort's goals is to support and lay the foundation for large-scale human space colonization, a rapid advancement of space power access that would revolutionize economic, political, and military space power.

This is not true. NASA has always been focused on political power from space. In the Cold War, NASA was the forefront of America's effort to win the space race, a political-technological showdown with the Soviet Union to decide whether the communist system or the free democratic system could perform the greatest technological miracles and win the popularity of the Third World countries in a play for favoritism. This political focus is unchanged today. NASA missions mostly are determined by political patronage and money sent to political districts by popular elected officials. The retirement of the space shuttle, planned for many years, is continually stalled by senators and congressmen whose districts are centers for space shuttle activity. The Constellation launcher rocket project is characteristically supported by politicians from the districts that will gain jobs and money from the program. None of these space projects is really examined through a space development lens by any of their political supporters. The civil space program is, and has always been, dictated by political interests which have never placed high emphasis on space development. This is not necessarily bad or wrong, but it does place at odds the actions of NASA versus the expectations of the people with regard to what NASA does. This dissonance has been a large factor in why NASA doesn't have the support of much of the citizenry (who believe they should be focused on space development), and is detested by much of the space advocacy community (whose visions include large-scale human spaceflight).

THE SPACE COMMANDS

Many will be surprised by this revelation, but the most well-funded and largest government space program in the United States is the military space program. Housed in the Army, Air Force, and Navy Space Commands (as well as smaller organizations such as the Defense Advanced Research Projects Agency), the total military space effort is numerically large and relatively highly impacting to society. Though little known or appreciated, it is this military program that most directly affects daily life. The Global Positioning System (GPS) satellite constellation operating by the United States Air Force Space Command provides precision navigation to anyone in the world with a civilian or military receiver, widespread equipment built by many countries besides the United States. Communications, imagery, and weather satellites that provide countless businesses and individuals with essential information services were pioneered by military space efforts. The space commands specialize in practical space technology because their mission is to support combat and other military operations through space technology, focusing on the military logic point, a most practical commission.

However, like NASA, developing general space power is not part of the space commands' commission. In fact, since the military space effort is parsed to the different land, sea, and air services, there is very little to no ownership of space as a distinct and valuable medium in itself. The space commands are interested in using current and shortly anticipated space technologies to advance their core missions. Air Force space systems are meant to support classic air missions, and similar arrangements are conducted by the Navy and Army. No efforts, beyond the writing and research of some military academics and strategists, attempt to advance space development or understand the complete utility of space efforts to national power. Space development in the space advocate sense is not considered a profitable activity for military space personnel.

Therefore, space development does not seem to be a high priority for either of the two main governmental agencies dedicated to space activities. This is probably the main reason that space development has not taken place in any meaningful way (beyond the success story of the commercial space industry, which is very good at expanding space technologies and initiatives pioneered by the government agencies) expected by popular imagination. Even though little space development has taken place, many visions of space development have been proposed in the years since the beginning of the Space Age. By looking at them we may be able to understand how to fix the serious space problem of lack of development in the "Final Frontier."

Janus and the Schism

The most important facet of the government space program today is that it is segmented into two distinct parts: a civilian "exploration" agency named the National Aeronautics and Space Administration (NASA), and a military "defense" sector composed of the individual services' space commands but concentrated in Air Force Space Command (AFSPC). Many see this arrangement as essentially correct. Exploration should be done by civilians, and the military should be kept out of space in all but only the most basic ways that space equipment can enhance terrestrial military power. However, this dichotomy was created almost out of whole cloth in the 1950s and early 1960s. Never before in American history has exploration, exploitation, and defense interests ever been bisected in such a way.

The American government's space program today can be likened to the Roman god Janus. Janus (the namesake of the month of January) was considered in ancient mythology as the god of gates or doors. He was often depicted as a single head with two faces, looking in opposite directions—gazing upon the future and the past simultaneously. In Janus, the space program finds an almost perfect representation: the master of the gates (to a prosperous human future in space) is a single head (the space effort at large) with two faces (the civilian and military programs) looking in opposite directions (in NASA, the future; in the military, the past). But here we will add another twist. Because Janus's vision is split, looking in two opposite directions, the head cannot make any significant progress in any direction and, as long as Janus does not unify his vision, the gates to space are permanently closed.

President Eisenhower's decision to create NASA was at least partially attributable to his disdain for some of the more outlandish statements about the future of the military in space made by the military, often Air Force officers, in the early years of the Space Age. Indeed, historians believe that the Eisenhower administration had a complete "lack of tolerance" regarding open speculation of potential military space missions and strategy.[13] James Killian wrote that Eisenhower administration scientists "felt compelled to ridicule the occasional wild-blue-yonder proposals by a few air force officers for the exploitation of space for military purposes.... These officers, often more romantic than scientific, made proposals that indicated an extraordinary ignorance of Newtonian mechanics, and the PSAC [President's Science Advisory Committee] made clear to the president the inappropriateness of these proposals."[14] Dr. Lee A. DuBridge, president of the California Institute of Technology and an Eisenhower PSAC member, went even farther, saying, "In many cases it will be found that a man contributes nothing or very little to what could be done with instruments alone," presaging an argument that has stymied the development of space for years.[15]

Killian and DuBridge were responding to quotes from Air Force leaders such as Brigadier General Homer Boushey, who mused on 28 January 1958, "The moon provides a retaliation base of unequaled advantage.... It has been said that 'He who controls the moon, controls the earth.' Our planners must carefully evaluate this statement, for, if true (and I for one think it is), then the United States must control the moon." While testifying to Congress in March of the same year, Air Force deputy chief of staff for development declared that a military base on the moon was "only a first step toward stations on planets far more distant from which control over the moon might be exercised."[16] The Air Force was not the only player in grand designs on space. The U.S. Army's Project Horizon study for a permanently manned (12–20 full-time residents) lunar outpost was completed in 1959 and contemplated a full moon base by 1966! Certainly these military programs were hopelessly exaggerated, right?

DuBridge railed against these studies as "wild programs of Buck Rogers stunts and insane pseudo-military expeditions." Killian by his own admission "ridiculed" the military's proposals rather than engage them on their merits in a civilized fashion. Killian, being president of the Massachusetts Institute of Technology, called these military proposals "romantic" rather than "scientific" in a blatant use of the logical fallacy known as "appeal to authority." But how are we to take this? General Boushey was the Air Force's Astronautics Division commander at the time of his moon base statement. The Project Horizon study was completed by von Braun's own Army Ballistic Missile Agency, which eventually formed the core of the Marshall Space Flight Center, which took man to the Moon in Apollo! These military space pioneers didn't just fall off the turnip truck, and the future Apollo engineers certainly knew Newtonian mechanics.

Perhaps the key to understanding the ridicule can be found in DuBridge's statement that man contributes nothing or very little to what can be done with instruments alone. Later in this same speech, DuBridge commented on the lunar base ideas, asserting that they were not necessary because "it is clearly easier, cheaper, faster, more certain, more accurate to transport a warhead from a base in the United States to an enemy target on the other side of the earth than to take the same warhead ... and shoot it back from the moon."[17] No doubt DuBridge's statement is correct. Eisenhower decided on 20 August 1958 to award the human spaceflight mission to NASA rather than the military partially for logic such as DuBridge's that declared there was no clear military justification for putting humans in orbit.[18]

Let us think of this logic chain in a different manner. Killian and DuBridge used the narrow vision of space that we identified above as the "Saganite" model of space development. These scientists were undoubtedly thinking of the scientific possibilities of unmanned satellite space research only, but instead of disclosing this narrow view of space's utility they instead ridiculed military plans as being "pseudoscientific" and "romantic" as opposed to their apparently superior "scientific" viewpoint. These scientists convinced Eisenhower that the military grand designs on space were products of inter-service rivalry only and military budget posturing. However, let's explore the military's motivations in their own words.

The Project Horizon study was very frank regarding the many unknowns in its estimation of the possibilities of the lunar base. In the report's section entitled "Background of Requirement," it stated:

> The full extent of the military potential cannot be predicted, but it is probable that the observation of the earth and space vehicles from the moon will prove to be highly advantageous.... The employment of moon-based weapons systems against earth or space targets may prove feasible and desirable.... The scientific

advantages are equally difficult to predict but are highly promising.... Perhaps the most promising scientific advantage is the usefulness of a moon base for further explorations into space. Materials on the moon itself may prove to be valuable and commercially exploitable.[19]

No doubt these statements were not iron-clad proof that a manned moon base was needed (especially to the scientists on Eisenhower's staff) but the Army specifically addressed exploring the enticing but still unknown possibilities for military, scientific, and economic advantage. Instead of enslaving the entire space program to the needs of "science" as the scientists wanted (surprise, surprise), the Army believed that its program could achieve national advantage along the national power spectrum known today as DIME (diplomatic, informational, military, and economic power). But in reality, the Army just envisioned the moon as a new frontier, and its Horizon base would serve the same purpose as its western forts during the U.S. expansion to the Pacific: protect settlers, miners, and the country from harm of the elements and hostile forces. In Project Horizon, the U.S. Army anticipated the O'Neillian/ Grahamian vision of space development. The military's so-called romantic statements may also have been more than simple grandstanding as the scientists believed.

The military's grandiose plans and speculations were not the foolish notions and romantic nonsense that Eisenhower's science advisors derided them as. In fact, these plans and statements were indeed the first attempts to try to define the nature of space activities in a national strategic context—the beginnings of space power thought. It's somewhat surprising and saddening that, as highly regarded a military man as Eisenhower was, he didn't understand that these musings were the natural (and commendable) reaction of the military attempting to come to terms with a new technology. It's doubly surprising that Eisenhower reacted so strongly against military space study considering that Eisenhower was an Army innovator as an early Army tank proponent when armor tactics were frowned upon by senior officers. His commitment to military modernization was also evident in his championing as president of an independent U.S. Air Force. Why, then, did he hate space thought so? Was it that he was already convinced the only utility of space was free overflight available to reconnaissance satellites? If so, his space vision was lamentably myopic. Was it his desire to keep federal spending reined in? Possibly, though his efforts were rewarded with the sour vintage of an unimaginably expensive space race with the Soviets which provided, not a lunar base or robust human spaceflight capability, but a "flags and footprints" stunt that has made the 1960s the high-water mark of human spaceflight even in the second decade of the 21st century. What is known is that the American space effort has not succeeded in opening robust space power development among all lines of power—economic, political, military—in ways most space enthusiasts have envisioned. By looking at episodes of exploration and development in American history, we find that perhaps a military-led O'Neillian/Grahamian space power vision may have better laid the foundation for true space power development.

Traditional Exploration and Development in the United States

Modern Americans tend to agree that the purpose of the military is to "kill people and break things." No doubt warfare is a very large and important responsibility of the military in wartime. However, this widespread belief has an often unspoken corollary that believes a military in peacetime is useless and irrelevant. It would surprise many people to know that this belief is a very modern, and very incorrect, view of the military in American history. In the great expeditions in American history prior to the second half of the 20th century, it was the military that was the forefront of exploration. Both the Army and Navy took part in great adventures taming both North America and far-flung foreign shores. In doing so, they provided great service to the country, and certainly by more constructive activities than simply killing people and breaking things.

When an American thinks "exploration," he most likely thinks of Lewis and Clark. What is less well-known is that what we know as the "Lewis and Clark" expedition was named by President Thomas Jefferson the Army's "Corps of Discovery" expedition that was led by military captains Meriwether Lewis and William Clark, and this expedition was only the first of many military exploration expeditions. In fact, from 1803 to the late 1870s, the U.S. Army's Corps of Topological Engineers (many officers of which were trained at the United States Military Academy at West Point) were the lead explorers of the American West and included such illustrious names as Captain (later General) Zebulon Pike (of Pikes Peak fame in Colorado Springs, Colorado) and the "Great Pathfinder" and legend of California, Major General John C. Frémont.[20] U.S. Army, not simply U.S. government, operations *exploring* the West led directly to its eventual colonization. Separate entirely from the Army's regular Corps of Engineers, the Corps of Topographical Engineers (authorized 4 July 1838) were officers dedicated to exploration, surveying, and mapping new lands and designing lighthouses and other aids to navigation. Historian James Ronda said the officers of the Corps of Topographical Engineers, "educated at West Point and connected to the latest in European and American scientific thinking, saw themselves as representatives of an expansionist nation as well as the larger empire of the mind."[21] Of the army explorer's 19th-century legacy, Ronda continues, "To brand the soldier-explorers as simply the advance guard for what George M. Wheeler called 'the ever restless surging tide of the population' makes for all too simple a story. The journeys of army explorers marked the foundations of a western empire. Those same journeys expanded the empire of the mind as well.... Soldier-explorers enlarged the boundaries of the American mind, making the West part of the nation's intellectual as well as geographic domain."[22] Army explorers, not simply explorers, made the West the American West—a far greater legacy than just killing people and breaking things.

The Army was not alone in the glories of American exploration. The Navy has its own proud heritage of exploration. The United States Exploring Expedition (U.S. Ex. Ex.) of 1838 took six U.S. Navy vessels and 346 men—including nine scientists—around the world in one of the greatest voyages of discovery in history. Only Chinese Admiral Zheng He's 15th-century expedition was larger. Historian Nathaniel Philbrick in his excellent book *Sea of Glory* says it all:

> By any measure, the achievements of the Expedition would be extraordinary. After four years at sea, after losing two ships and twenty-eight officers and men, the Expedition logged 87,000 miles, surveyed 280 Pacific islands, and created 180 charts—some of which were still being used as late as World War II. The Expedition also mapped 800 miles of coastline in the Pacific Northwest and 1,500 miles of the icebound Antarctic coast. Just as important would be its contribution to the rise of science in America. The thousands of specimens and artifacts amassed by the Expedition's scientists would become the foundation of the collections of the Smithsonian Institution. Indeed, without the Ex. Ex., there might never have been a national museum in Washington, D.C. The U.S. Botanic Garden, the U.S. Hydrographic Office, and the Naval Observatory all owe their existence, in varying degrees, to the Expedition.
>
> Any one of these accomplishments would have been noteworthy. Taken together, they represent a national achievement on the order of the building of the Transcontinental Railroad and the Panama Canal. But if these wonders of technology and human resolve have become part of America's legendary past, the U.S. Exploring Expedition has been largely forgotten.[23]

This extraordinary feat of logistics, exploration, and professionalism is a credit not only to the United States, but also to the United States Navy. But one must pause and consider why both the military character of the exploration of the western United States and almost the entirety of the greatest modern sea exploration endeavor in history is completely forgotten or ignored today. I find it hard not to believe that this is partially (if not wholly due) to a modern preoccupation with dismissing the positive qualities of the military and desire to eliminate the institution by misguided utopian idealists, but it's up to the reader to decide.

Regardless, to deny the military's proud history as the *preeminent* exploration force in American history is to make a profound and inexcusable error. Yet, in the early Space Age that exact heresy was done. History indicates that Eisenhower's odd distrust of military motives (and equally damaging overregard for civilian scientist motivations) robbed the Air Force of its own great contribution to American exploration. Air Force Brigadier General Simon "Pete" Worden argued for the Air Force to reassert its role in exploration in his 2002 monograph (with Major John E. Shaw) *Whither Space Power? Forging a Strategy for the New Century*:

> Currently, NASA is the agent for U.S. space exploration. This is counter to traditional American approaches to exploring and exploiting new territory. It is also counter to common sense. NASA is a research and technology organization. It has little incentive to develop, open, and protect new areas for commercial exploitation.[24]

General Worden certainly knows what he's talking about. He was responsible for perhaps the 1990s' greatest exploration mission to the Earth's moon—the *Clementine I* satellite project. This little satellite produced the first full, multicolored map of the Moon in 1994 using new low-cost technology developed for use in missile defense programs. Moreover, it was a Defense Department Strategic Defense Initiative Organization, not a NASA project, needing only two years to develop and $80 million dollars to complete.[25] This mission proved that the military could still explore just as well—if not better—than the competition. Of significant interest is that the mission paid strict attention to the commercial viability of lunar operations by "prospecting" the Moon's resources via surface imaging. It was nicknamed "the Miner's daughter," recalling the military's historical recognition of economic expansion as a national priority. Worden continues his musings on the future of military space exploration:

> [T]he defense establishment's foray into space exploration is only beginning, not ending. Strictly "scientific" exploration of space or any other new area is quite acceptable as long as nothing of value is seen in it and no threat can emerge from the new territory. This was the case for the Antarctic continent and has certainly been the case for objects beyond [geosynchronous Earth orbits]. However, just as the global economy is migrating to increased reliance on [Earth orbiting] global utilities, so there will be both potential threats to them and economic resources to protect. The same will inevitably occur with the Moon, asteroids, and other objects

in the solar system.... *The military services will inevitably return to their traditional roles: Protect commerce, deny access to adversaries, and discover new resources.*[26]

Did Eisenhower's scientists see "nothing of value" in space or ridicule the military's interest in space because they wanted to keep the space program for themselves? It would appear that this is the Saganite motivation. The military in the '50s and '60s may not have seen exactly what was of value in space, but they *knew* it was there and were willing to expend a great deal of time, energy, and resources to find it. As we will argue later, they would have found space's true value sooner or later, and long before the narrow vision of scientists would have allowed.

Now that we are finding the value of space, and that it is the military finding it, could we have found and exploited this value earlier if we had allowed the military to perform its traditional function in space without stripping it of its role with a civilian "exploration" agency staffed by narrow-minded scientists? Air Force space pioneer General Bernard Schriever said of NASA's creation that he "was very much opposed to the organizational arrangements right from the beginning. NACA [the National Advisory Council on Aeronautics, the military-friendly precursor to NASA] should never have been disturbed. Creating NASA was an unnecessary creating of an organization.... [The government] simply took the military, put them over in NASA and started the manned space flight program. They would've done much better had they allowed the military to carry out the operational type of flying. We proved that we could do it. We had our people running the programs. Eisenhower was sold a bill of goods by [science advisor] Jim Killian."[27] To this, General Schriever could have added that the military's historical broad view of national defense responsibilities would have allowed for a much broader view of "valuable" activities in space as well.

What must be considered by historians and space theorists alike is the possibility that only in the military was the institutional history, vision, and logistical expertise necessary for rewarding (across all aspects of national power besides simple "science") and sustainable space exploration possible, and that civilians in the form of NASA did not have the necessary breadth of vision to chart humanity's push into space. Put simply, science benefits alone could not justify space travel and so NASA was pushed into the base field of politics and bureaucracy. The Saganite vision of scientists was chosen based on short-term myopia and an irrational elevation of scientists in over the grander and more rewarding proto–O'Neillian whole-system visions of the military, with disastrous results to space development that still reverberate as lamentable lost opportunities over six decades later. But was the Air Force actually showing signs that it was looking in a better direction than NASA?

USAF Man in Space

Contra to President Eisenhower's "Space for Peace" platform, the Air Force believed that military activity in space could and should be considered peaceful. Air Force activity in space was no different than the peaceful activities of the U.S. Navy on the high seas—assuring access to the oceans of all parties engaged in peaceful activity.[28] Especially essential to Air Force activities was extended exploration of the utility manned spacecraft could have in space. The lack of human access to space is considered by many space advocates to be a key reason for the perceived lack of progress in space. Many blame NASA's insistence on holding its monopoly on human space travel and its emphasis on "science" over "exploitation" as reasons for a human space travel capability adrift. What should not be ignored is that human spaceflight was *the* Air Force space policy priority at the beginning of the Space Age.

Strategic Air Command commander General Thomas Power stated the Air Force position on human spaceflight as, "For the long term, the critical requirement is to establish man in the space environment. In the early-unmanned exploratory stages of the conquest of space, unmanned vehicles can be used for many scientific purposes, and certain specific military applications. However, to fully exploit the medium, man must be the essential ingredient."[29] Very few space advocates today would disagree with General Power's statement. General Schriever amplified the Air Force's position in a November 1961 document entitled *Manned Operational Capability in Space*, which stated:

> The best approach to our military space program is a mixture of unmanned and manned space vehicles. More emphasis on manned spacecraft is required. We must be able to use space on a routine, day-to-day basis.... Man's abilities [in space] are necessary to support our national objectives and national security in the space age.... Finally, the key to rapid utilization of space by man for military or civilian purposes is flexibility. *We must not design our space vehicles and programs just to achieve those objectives which we can define now. We must design the vehicles with enough capacity to rapidly adapt to or incorporate the vast new knowledge which will flow from our space program.... Historically, we have tended to overestimate what we could do on a short-term basis and to grossly underestimate what we could do on a long-term basis.*[30]

A case can be made that the Air Force's interest in manned spaceflight was solely a selfish interest in claiming a large piece of the budgetary pie and narrow interests in defending the future of its "flight, fight, and win" culture. There was institutional concern that the Air Force officer corps would change "dashing and courageous pilots" into nothing but "silo-sitters" (which, incidentally, is the modern fate of U.S. Air Force space forces). The Air Force wanted a follow-on to the fighter for its pilots, and many saw space as footing the bill should flying ever drop off.[31] Historians seem to have concluded that this explanation is correct, and that the Air Force used these arguments because they could not convincingly articulate and precisely what humans would do in space "to the satisfaction of its civilian overseers" and thus failed to establish an independent, long-term, human presence in space.[32]

I submit that these arguments made by the Air Force were not the last gasps of a defeated foe trying desperately to keep its programs alive, but are in fact the correct arguments given the situation. These were not pie-in-the-sky comments, they were the wise words of military institutional experience with a long experience in evaluating new technologies. There is simply no way to know what value a mature space operational capability could be so early in the Space Age, and the most honest and truthful answers the military could give is to say they needed to experiment in space before they could be certain, but they sense the advantages could be spectacular! The unknown is the way it is because it is unknown. There is no way Orville and Wilbur Wright could have foreseen a heavy bomber or supersonic transport plane on Kitty Hawk, and it would have been outrageous to demand of them to do so before embarking on building a plane. H.L. Hunley, designer of the world's first successful attack submarine, could never have imagined the cat-and-mouse games on the U.S. versus the Soviet "silent services" played by deadly assassin attack submarines and their ballistic missile submarine prey. To demand instant gratification from a new technology is often simply impossible, and incredibly simple-minded.

Consider the observation of Eisenhower's second science advisor, George Kistiakowsky, that the military's grandiose military space projects "were quite partisan, to put it mildly.... Rather awful! ... I still recall becoming indignant on discovering that the cost of *exclusively paper studies* in industrial establishments on 'Strategic Defense of Cis-Lunar Space' and similar topics amounted to more dollars than all of the funds available to the [National Science Foundation] for the support of research in chemistry."[33] If true, was military expenditure on studies excessive? Perhaps. But is it really surprising to hear from a *scientist* that *science* wasn't being adequately funded compared to other priorities? Look at Kistiakowsky's derision of "exclusively paper studies." In military circles, these paper studies are known as explorations into the art and science of warfare and are very important—modern navies would not exist today if not for Admiral Alfred Thayer Mahan's 19th-century book *The Influence of Sea Power Upon History*, certainly a very important exclusively paper study. Today, there is still no widely accepted consensus on a theory for space operations, known as space power theory. Why should we believe this scientist is correct when it is obvious he knows nothing of the importance of military strategic research? Therefore, in part because we listened to scientists ignorant of military operations, we spend a great deal of money in space without having an overriding strategy of what is best to accomplish in space. Many scientists and even some military professionals conclude that because there is no space power theory that explains why space operations is important, then space operations simply isn't important (an argument often used to criticize calls for an independent military space service). These people forget that absence of evidence for something is not evidence of absence.[34]

Eisenhower, beyond simply ignoring the importance of military theoretical work and deriding good officers merely doing their job as they understood it, pressed the space effort in another very important and damaging way: the pocketbook. More specifically, he did not treat space power as a new and promising field of endeavor in itself, but merely as an adjunct to try to do terrestrial military operations better—a philosophy that has led to the near-total blindness of the military to the possibilities of space-focused military power today. Eisenhower's approach to space finance was to avoid duplication, wasteful expenditures, and overlap among space programs.[35] This is a laudable goal in theory, but in reality bureaucrats and scientists (already proven to be disdainful and jealous of the nation's interest in space) not space experts are ill-equipped to understand what truly constituted wasteful overlap and duplication. Scientists tended to see that "man in space" was a requirement. NASA's Mercury program was a man in space. The USAF's Dyna-Soar manned military spaceplane was also a man in space. Therefore, the two programs are needless duplication. Dyna-Soar must be cancelled. So goes the line of thinking. But where Mercury truly was little more than "spam in a can," the Dyna-Soar was a very advanced and capable craft that could have had many different uses both military and scientific. Perhaps only a space power expert could have seen the difference (I believe any normal citizen could have seen a massive difference and would have been far more impressed with the Dyna-Soar), but a hostile scientist could be dismissive enough to say "wasteful overlap" with a straight face.

Eisenhower's philosophy of military space funding can be described as "unless there was a functional efficiency of performing a terrestrial military mission in space, or it was cheaper to perform a military function in space than on Earth, no space activity would be funded." Little exploratory research and development would be authorized.[36] Essentially, space had to pay off in the short term or there would be no space effort. Anyone familiar with the concept of basic versus applied research and development knows that without basic R&D exploring the unknowns to determine what is possible, applied R&D into taking advantage of the possible dries up. Basic R&D is the applied

R&D a decade later. Eisenhower's insistence on only performing traditional Earth military operations in space was a severe blow to military theoretical basic R&D in space power thinking. Essentially, Eisenhower prevented military minds from being supported in thinking about how to approach space for national strategic benefit. No wonder that during the Eisenhower administration "[space] technology [had] far outpaced any coherent doctrine on how to employ space systems effectively."[37]

Even with Eisenhower, dazzled by his science advisors, resisting military attempts to explore space power theory—an eminently logical, useful, and traditional military endeavor—there is evidence that the Air Force got incredibly close to developing a consistent space power model presaging the O'Neillian space development vision that would have placed the United States and humanity on a solid footing for conquering the space environment in ways NASA has only dreamed of. Unfortunately, strategic culture counts, and it appears that the Air Force was ill equipped to answer serious space power questions because of its incompatible experience flying aircraft in World War II. However, the Air Force was asking the right questions, and given time, would have stumbled onto the correct answers to make the United States a true spacefaring nation. Let us turn to the Air Force's wrong turn in justifying military space activity and look at a program that would have ultimately created the Starfleet future as early as the late 1950s.

The Air Force's Wrong Turn

Strategic culture is a fact of life in the military. A military service's strategic culture is essentially what the service thinks it should do both in war and peace, and how it should be organized, to best protect the country. In many ways, the service culture is based on the strategic power theory that it most identifies with. Among the strengths of having a clear and powerful service culture is that the service's members know what their role is in the larger structure of national defense, and the service's thinkers are confident enough to remember the past as well as use their command of history to project activities into the future. A significant negative of service culture is that it is sometimes so strong that when confronted with something so different from their experience the original culture might not be able to address it correctly and must yield to a new interpretation of a service culture or bungle its new responsibilities.

The Air Force's strategic culture was the one that confronted space the most during the early Space Age. As we have seen, the Air Force's strategic culture remembered its military culture roots (steeped in historical tradition of exploring for military and commercial gain) enough to recommend the exploration of space technology for advances among all levels of national power: diplomatic, informational (including scientific), military, and economic—advocating an O'Neillian vision for a large and diverse human space effort counter to Eisenhower's science advisors' limited and self-centered Saganite view of space as a playground/preserve for scientists only. Space enthusiasts would find much in Air Force and military space thinking that they would agree with, and life a half century later would be far different had we moved in the military, rather than the "scientistic," direction.

However, the Air Force service culture was not perfect by any stretch of the imagination. It had a fatal flaw that did not and maybe could not have endeared itself to rational decision makers. This flaw was due to the fact that the particulars of the Air Force's strategic culture are completely incompatible with space power and interpreting space power through an Air Force lens in the age of nuclear bombardment and Strategic Air Command doomed contemporary military ruminations of space power to look to civilians like the rantings of bloodthirsty and psychopathic lunatics, even though Air Force statements were actually just standard Air Force thoughts applied to space. And even if these thoughts were odd, there is ample evidence that they would have been abandoned quickly as the Air Force gained experience in space and would have began to form a space strategic culture more attuned to the requirements of national and economic security in space. The problem, essentially, is the Air Force's lessons learned from World War II.

Ever since the first Army airman dreamed of what could be done with the Wright Flyer, military thought on air power has been obsessed with the offensive. Virtually all of the important air power thinkers prior to World War II experienced the grotesque, stalemated trench warfare of World War I and vowed to use the airplane to never allow it to happen again. Thoughts immediately converged to conceptualize heavy bombers—large aircraft capable of flying over the stalled and immobile ground forces and strike targets deep in the enemy interior to weaken enemy resolve and sue for a quick peace. Specific tactics ranged from the rather extreme methods of Italian theorist General Giulio Douhet, who advocated dropping poison gas bombs on enemy cities and specifically target civilian populations to the more tame American Brigadier General William "Billy" Mitchell's desire to use bombers to strike industrial centers that could cripple and paralyze an enemy's ability to equip and sustain its military force in the field, causing systemic collapse. That is to say, there wasn't much range at all. There was quick and early consensus in air power thought that an enemy's "vital centers" needed to be struck, and the only real debate was whether civilian casualties should be desired specifically or merely minimized to acceptable amounts when targeting industrial targets.

In World War II, technology and a vicious war allowed air power to reach maturity. Both Douhet's and Mitchell's theories were tested and applied. The results were hundreds of thousands of civilians killed in air attacks. Mass Allied firebombings of Dresden and Hamburg (thought to have killed hundreds of thousands of people at first, but now regarded to have killed 25–50,000 people each) and Tokyo (75–200,000 killed) were equal to or greater in ferocity than the atomic strikes on Hiroshima and Nagasaki (total casualties may have ranged up to 200,000, taking into consideration long-term effects). Whether the military results of the bombings justified these deaths is heavily debated, but what is important is that air power was judged to be successful during the war, and air power became justified through its use of heavy bombing. The Air Force's strategic culture became long-range heavy bombing to destroy an enemy's ability and will to fight.

Enter the Air Force during the Space Age. The Strategic Air Command reigned supreme. SAC's fleet of manned bombers provided an insurmountable deterrent to Soviet aggression (known as deterrence) by retaining the ability to reduce the Soviet Union to a radioactive cinder within hours should the president (or perhaps even SAC generals) deem it necessary. Since the Air Force strategic culture embraced massive destruction through aerial bombardment, it stands to reason that Air Force spokesmen would first believe that massive bombardment from space would be the ultimate use of military space power. It was simply in human nature to view a new environment in terms of the environment you are most familiar with. In this light, it seems natural that SAC General Power would speculate on space operations as simple extensions of what the Air Force already did in the air:

> We must not, in the fashion of decadent nations, permit our gross potential to be bled off into purely defensive weapons. As we enter the space era the primacy of offense has never been more clearly defined.... Because space offers the ultimate in mobility and dispersal for weapons which can be addressed at the enemy heartland, the ultimate in deterrence may well be in this direction.... [The Air Force must] emphasize constantly the positive contribution of offensive weapons systems. The logic of this fact must be identified for scientific and national leaders.[38]

Here Power has perfectly addressed performing Air Force missions in space. To Air Force leaders, the offensive weapon was most powerful, deterrence was their mission, and space may well be the ultimate environment to pursue perfect deterrence. This is a clear example of Air Force strategic culture seeing itself in space. To the Air Force, this was simple logic. To civilians, it sounded like turning the heavens into the ultimate field to plot humanity's extinction. Perhaps it was both. But even with this bloodthirsty demeanor, the Air Force envisioned much more than deterrence in the face of nuclear Armageddon in space. General Boushey was a bit more inclusive in his thoughts on the ultimate expression of Air Force space thought:

> In twenty years [1979], I believe both the moon and Mars will have permanent, manned outposts.... Another use [of satellites] will be purely military—bombardment—and accomplished by space vehicles. I use the term vehicles rather than satellites because I believe these systems will be manned.... It appears logical to assume we will have antisatellite weapons and space fighters.... [The only thing that will cost more than these space systems] would be the failure to be the first on the moon. We cannot afford to come out second in a territorial race of this magnitude.... This outpost, under our control, would be the best possible guarantee that all of space will indeed be preserved for the peaceful purposes of man.[39]

Even though Boushey highlights the Air Force cultural concepts of fighters and the bombardment mission in space, it is clear his strategic vision sees more than just higher-altitude bombers and fighters. Embedded in his proposal are space colonies (in the form of bases) and territorial expansion (often regarded as necessary by space enthusiasts today) and making space safe for all "peaceful purposes of man," not simply scientific or military endeavors. Even so, it is unsurprising that civilians hearing this and other quotes surmised that all the Air Force wanted to do was expand SAC and nuclear weapons into space. With this belief, it is quite understandable that civilian policymakers would turn to a civilian exploration agency to keep the military from making space the ultimate offensive weapon. However, a little-known Air Force plan for the ultimate nuclear weapon in space, if built, may have actually made nuclear weapons far less dangerous *and* propelled the United States into a much more beneficial space power future and correct space power path almost immediately upon entry into the Space Age. Even the offensive-minded and somewhat scary Air Force mind may have understood enough of the truth of space power theory to give birth to a truly great and beneficial space future through the construction of the *Orion*-class space battleship and the development of the Deep Space Force.

Deep Space Force: Access and Ability into the Solar System

Eminent science fiction author and space visionary Sir Arthur C. Clarke called the Orion project "one of the most awesome 'might have beens' of the space age." *Orion* was envisioned as a manned spacecraft propelled by the shock waves from miniature hydrogen bombs. Literally, miniature hydrogen bombs would be pushed out the back of the spacecraft and the resulting explosion would push the spacecraft forward at great velocities as it rode the shockwaves of atomic explosions. In essence, this "nuclear pulse" engine would achieve the "Holy Grail" of spaceflight—provide a space travel engine that would be very fast, very efficient, and high thrust at the same time. To this day, it is the most powerful space engine designed that can be developed into an operational spaceship (ignoring, for a moment, the very real debate over how safe this engine would be to use!). From 1957 to 1965, physicists (many veterans of the Manhattan Project) and Air Force officers developed and studied this remarkable craft. In a time when the specific impulse (or ISP—a measure of rocket efficiency) of the best chemical rockets only reached about 400 seconds, the Orion vehicle offered 2–3,000 seconds at first, which advanced designs yielding 6,000 seconds. Instead of struggling to send two astronauts to the Moon for a few days in 1969 as with Apollo, Orion could send a real spaceship with more than a dozen crewmembers to Mars in 1965 and continue with Project Orion's unofficial motto: "Saturn by 1970!"[40] For interested readers, Orion's remarkable story is told by Orion scientist Freeman Dyson's son George in his remarkable book *Project Orion: The True Story of the Atomic Spaceship*. For our purposes, we'll discuss one of the proposed missions of a military Orion spaceship and the serious proposal for the creation of the Deep Space Force.

"Although the *Orion* propulsion device embraces a very interesting theoretical concept, it appears to suffer from such major research and development problems that it would not successfully compete for support." With this statements, NASA administrator Richard Homer rejected space agency support for the *Orion* concept.[41] It would be up to the Air Force and the military alone to continue to develop the most advanced and powerful space propulsion system yet devised by human intelligence. However, even if the military had a broader view of the potential of space activities to benefit human civilization, the Air Force still needed a reason to support the *Orion* vehicle for military purposes. True to the Air Force's strategic culture, the military purpose of the *Orion* would be to bolster the Air Force's primary function: America's nuclear deterrent.

In 1960, the world's stockpile of nuclear weapons was estimated at 30 million kilotons.[42] However, these forces in the form of manned bombers and strategic missiles were always on hair-trigger alerts. They had to be, for in half an hour a "first strike" nuclear assault would reach its enemy and completely destroy the target's country. The nuclear deterrent had to be able to respond in under 30 minutes or it would be destroyed in the assault and be virtually worthless. Faced with this reality, both American and Soviet militaries enacted very strict and complicated systems that ultimately kept the peace of the Cold War, but frightened even their most ardent supporters because one unantici-pated failure of the system of either side could result in the complete destruc-tion of the human race. Even the most ardent SAC supporter was all ears to find a safer alternative: Enter *Orion*.

Air Force Captain Donald Mixson, a physicist working on Project Orion at the Air Force's Special Weapons Center at Kirtland Air Force Base, New Mexico, offered a Deep Space Force comprised of military Orion spaceships as a saner alternative:

> Once [an Orion] space ship is deployed into orbit it would remain there for the duration of its lifetime, say 15 to 20 years. Crews would be trained on the ground and deployed alternately, similar to the Blue and Gold team concept used for the [Navy ballistic missile] Polaris submarines. A crew of 20 to 30 would be accom-modated in each ship. An Earth–like shirt-sleeve environment with artificial gravity systems, together with ample sleeping accommodations and exercise and recreation equipment, would be provided in the space ship. Minor fabrication as well as limited module repair facilities would be provided on board.
>
> On the order of 20 ships would be deployed on a long-term basis. By deploying them in individual orbits in deep space, maximum security and warning can be obtained. At these altitudes, an enemy attack would require a day or more from launch to engagement. Assuming an enemy would find it necessary to attempt destruction of this force simultaneously with an attack on planetary [i.e., Earth] targets, initi-ation of an attack against the deep space force would provide the United States with a relatively long early warning of impending attack against its planetary forces.
>
> Each space ship would constitute a self-sufficient deep space base, provided with the means of defending itself, carrying out an assigned strike or strikes, assessing damage to the targets, and retargeting and restriking as appropriate. The spaceship can deorbit and depart on a hyperbolic Earth encounter trajectory. At the appro-priate time the [presumably nuclear] weapons can be ejected from the space ship with only minimum total impulse required to provide individual guidance. After ejection and separation of weapons, the space ship can maneuver to clear the Earth and return for damage assessment and possible restrikes, or continue its flight back to its station in deep space.
>
> *By placing the system on maneuvers, it would be possible to clearly indicate the United States' capability of retaliation without committing the force to offensive action. In fact, because of its remote station, the force would require on the order of 10 hours to carry out a strike, thereby providing a valid argument that such a force is useful as a retaliatory force only. This also provides insurance against an accidental attack which could not be recalled.*[43]

The benefit of this Deep Space Force to SAC is clearly stated in Mixson's last paragraph. SAC's motto was "Peace Is Our Profession" but this peace was upheld at quite literally a cocked gun at the head of the entire planet. With the Deep Space Force, overwhelming retaliation could be accomplished with a great deal of time to spare, making accidental un-recallable nuclear war scenarios such as in the movies *Dr. Strangelove* and *Fail-Safe* a nonissue. This Deep Space Force does sound quite fantastical in a way, though many of the 20th century's greatest scientists swear the ships could have been built. Dyson says of the Deep Space Force's military utility, "Was it crazy to imagine stationing nuclear weapons 250,000 miles deep in space? Or is it crazier to keep them within minutes of their targets here on Earth?"[44] We could also add, Would it have been crazy to station America's nuclear deterrent in spacecraft where a Soviet first strike against the U.S. nuclear forces would have killed 400 servicemen of the Deep Space Force? Or is it crazier to place America's nuclear deterrent near America's major cities where the Soviet first strike against it would also kill millions of citizens as collateral damage? Was the Deep Space Force such a bad idea?

Dyson speculates about other potential advantages of the Deep Space Force:

> The Blue and Gold Orion crews would have spent their tours of duty on six-month rotations beyond the Moon—listening to 8-track tapes, picking up broadcast television, and marking time by the sunrise progressing across the face of a distant Earth. With one eye on deep space and the other eye on Chicago and Semipalatinsk, the Orion fleet would have been ready not only to retaliate against the Soviet Union but to defend our planet, U.S. and U.S.S.R. alike, against impact by interplanetary debris.[45]

No doubt the Deep Space Force would have been ready when humanity woke up to the threat of asteroid and comet impacts in the late 1970s, but let us speculate further. The parallels of a Deep Space Force cruiser crew and a *Star Trek* starship crew are remarkable. Project Orion team member David Weiss even stated that Orion was being pitched to have multinational North Atlantic Treaty Organization (NATO) crews, making the Deep Space Force an international organization.[46] Also, a six-month cruise would allow for a great deal of down time for crews, and there is little doubt that in between weapons drills and routine maintenance on board a DSF cruiser there would be plenty of time for space science, exploration, and thinking about how best to use space for human benefit. People would likely conclude that there were other things worth doing besides Orion crews twiddling their thumbs waiting for World War III. Before long, DSF cruisers would probably have gained a dedicated science officer or staff section and one or more ships may have left the deterrence fleet to catalog potential hazardous asteroids or comets, provide emergency rescue services to commercial operations, map resources on the Moon, and perhaps even send a human fleet on a United States deep space exploration expedition scientific mission to Mars or Saturn, channeling the great military exploration expeditions of the past. What should be obvious is that the collateral value of a single Deep Space Force *Orion* vessel would likely be greater than the entire portfolio of NASA space equipment ever fielded! This includes functionally limited single-use manned spacecraft (*Mercury*, *Gemini*, *Apollo*, and space shuttle) and unmanned probes put together, and would not have been appreciably more expensive.

Eventually, Project Orion and the dream of the Deep Space Force died when the Air Force cut its funding because the Air Force could not carry Orion's financial burden alone.[47] It was simply too expensive and the military utility alone couldn't justify Orion's development cost. However, Project Orion's team members always considered the military utility of Orion to be a rationale for developing the ships only. The purpose of Orion was to explore. Captain Mixson's proposal for the Deep Space Force was written "not to make Orion a military machine, but to con a military machine into yet another installment of funds to keep [the] big beautiful dream" of space exploration alive.[48] Perhaps, if the military wasn't stripped of its historical responsibility for exploration early on in the Space Age, this "con" would have been unnecessary and the United States would have received the space program the innovative and visionary country deserved in Orion. Unfortunately, history and politics had different plans.

So here we have an example of Air Force strategic culture devising a military spacecraft that could have been used as the centerpiece for developing a robust space power operating philosophy. Dyson maintains that Mixson's inspiration for the Deep Space Force was Mahan's *The Influence of Sea Power upon History*. While not in the Air Force's strategic culture, Captain Mixson's familiarity with the larger military culture allowed him to borrow from naval strategic culture to forge the beginning of a unique space strategic culture. Mixson had the strategic understanding to see what should happen in space that Eisenhower science advisors Killian and Kistiakowsky did not. The American space effort is the worse for it.

The Aborted Space Age: Bitter Harvest of the Saganite

One can certainly dismiss this history. Civilian explorers such as NASA can do everything the military can do, but without the "evil" connotation baggage of the military. Perhaps, but the military's "bad" reputation is a 20th-century phenomenon and is a modern misconception. As space policy analyst Eileen Galloway eloquently stated on 11 May 1958:

The fact that one scientist wears a uniform while his co-worker wears a civilian suit does not mean that the uniformed scientist is an incipient Napoleon who threatens popular government.... Control by a group of scientific specialists is just as dangerous to a democratic government as control by a group of military specialists. [The important point is the] concept of control of policy by the elected representatives of the people over the various professional specialists who lack the breadth of vision required for guarding the common welfare and the public interest.[49]

The American people must understand that the military does not equal violence and senseless death. NASA is not "good" because it is civilian and the military space program or an independent military space service is "bad" because it is military. *Star Trek* clearly shows scientists in military uniforms as heroes. With this broad popular support and acceptance of military space activities in fiction, it is somewhat surprising that real military activity in space is frowned upon so greatly by those professing to be space enthusiasts today. The simple fact is that short-sighted civilian scientists in the 1950s led us on the path of the narrow Saganite vision of space development which has so retarded the space development expected in the modern imagination. The military alternative was clearly an O'Neillian/Grahamian type of space expansion through parallel development in civilian science and economic expansion and military missions. By wrongly ignoring the military's traditional role in exploration and expansion, dismissing the military's consideration of larger factors in space development than simply "science" as the prattling of ignorant children, and being irrationally convinced of the objectivity of scientists, the Eisenhower administration set the United States on a devastatingly inefficient and straitjacketed course in space that could not develop space even with the astronomical amounts of money that would be allotted to space activities in the Kennedy and Johnson administrations.

However, these policy limitations are self-inflicted wounds and are easily corrected. Perhaps the most important policy change that will allow the United States an efficient organization for space power logic will be to place the historically broadest thinking space power organization—the military—back into prominence with a new mandate to develop space power in all of its forms, and not simply focus on warfighting. What would the General Theory advise for such a rechartered military space organization?

Space Power Logic in Military Space Organization

The General Theory assumes that the best organization for military space is the one that allows the most development through the five paths applied to space power's logic and grammar. While grammar development (mostly technical research and development, the purview of military matériel commands) is an essential part of space power development, the most critical developmental responsibilities are Path 2 (new ways to handle space power elements) and Path 5 (new methods of organization) at the transformer level of space power logic. The military's most essential responsibility is to develop the transformers that can change raw space power into applied military space power. Creating military transformers is more commonly known as strategic and tactical development. Because developing space power grammar is a broad responsibility shared among government and private interests, the best military space organization is the one that can develop the best transformers through military space theory, strategy, and doctrine at the strategic and tactical levels. With a strong transformer base in space power logic, not only will applied military power be better than one without such a mature logic understanding, but through the transformer feedback loops, space power grammar development will be improved as well. However, before we can decide upon the best organization, we must determine the characteristics of the best organization itself.

An oft-used argument against a separate space force is the perceived dead weight cost of standing up an entirely new service: uniforms, bases, infrastructure like personnel, medical, support services, and administrative expenses that do not translate directly into increased operational power or capabilities. Admiral David Jeremiah, a 2001 Space Commission member, summarized this sentiment as the "tooth to tail ratio."[50] It was this low return for high overhead that caused the commission to believe a 2001 independent space service would be "dysfunctional."[51]

The "tooth" is presumably combat power and battlefield or national security effectiveness, while the "tail" is the overhead and administrative costs. In this framework, the sum total of a service's worth is its combat capability "tooth" while the "tail" should be as small as possible. Ideally, an infinitely long tooth would be attached to an infinitesimally small tail. However, is it true that combat power is the only measure of a service's worth?

A service's worth is more than the sum of its tangible elements: its effectiveness comes from, not simply its men and matériel, but also its strategic culture and its ability to think and add to the national defense discussion. The "tooth" is an incomplete measure of the service's usefulness. A better measure is rather the service's "head," the sum total of its combat and operational ability plus its contribution to what Admiral Wylie calls the "differences in judgment ... the clashes of ideas" and the "intellectual reserve, a reserve of strategic concept."[52] The strategic culture of the service is every bit as important as its weapons or operations at any one time.

The "head to tail" ratio has a very different calculus than "tooth to tail." "Tooth to tail" emphasizes immediate effect to the exclusion of context and growth. "Head to tail" is more inclusive, with context at its center in order to mature the strategic culture as well as the service's national defense capability. A military organization fights with both its tools and its head (its mediums grammar and logic). Therefore, the "tooth to tail" ratio is an inadequate tool to gage the utility of an independent space service. The strategic culture must be considered. Under "head to tail," we stand a much better chance of making the best decision on the question of an independent space culture and service.

Of course, because the organization's head is so important, it's critical to make sure the organization's head is in the right place. Put simply, some organizational cultures are better than others, even among organizations dedicated solely to space power. Holley again is useful with a historical example from air power history. Indeed, he even uses an example from a much ignored part of space power history!

When Lt Hap Arnold was groping tentatively into the unknown future of the aeroplane in 1913, the Army authorities already had decided to assign the aviation mission to the Signal Corps. What were the implications of taking that organizational turn in the road? The Signal Corps was *not* one of the combat arms; it was a *service*—one of the ancillary branches that render support to the combat arms. That decision, allocating aircraft to the Signal Corps, was to play a critical role in determining the future of the air arm for many years to come.

The organizational or institutional bias implicit in being a *service* seemed inexorably to warp the conception of the role aircraft were to play in the years ahead. As the principal agency for communication or the transfer of information, it was entirely natural for the Signal Corps to stress the support role of the airplane, the gathering of information—aerial photography, observation, reconnaissance. The airplane provided the eyes of the Army in a new and wonderfully enlarged way. Indeed, airplanes proved to be far better eyes, more versatile, faster, and with greater range, than any eyes the Army had ever had before...

There were a number of reasons why the Army gave primacy to observation and related close air support roles. One of the principal reasons lay in the fact that the Army lacked an adequate organization and method for the systematic analysis of its operational experience. It was, therefore, ill-equipped to develop a sound body of doctrine. Since the experience of the [American Expeditionary Force] with aviation, especially with strategic bombing, was exceedingly brief, deriving sound doctrine was a difficult task at best. So the chief of the Air Service simply mirrored the major body of experience, which was in observation, and failed to see the enormous potential hinted at in the limited body of experience with strategic bombing.

There was, of course, a very good reason for assigning aircraft to the Signal Corps. In 1909 that service was one of the most progressive, one of the most scientifically inclined of all the arms and services. Leaders in the Signal Corps ... were nationally respected for their contributions to science. But surely it would have made more sense *doctrinally* to assign aircraft to the Cavalry.

Reflect a moment on the traditional doctrinal roles of Cavalry as a combat arm. First, there was the long-range, deep penetration *strategic mission*—strike the enemy homeland, disrupt transportation and communications, and burn factories. Next, there was the *screening mission* using the speed differential of the horse as compared with marching men to fan out in front and on the flanks to give a tripwire against enemy approaches and to conceal friendly concentrations. Third, there was the *interdiction mission*—attacks against the flanks of enemy columns before they can close with the friendly main battle force. Fourth, there was the *reconnaissance role*—serving as the eyes of the army, giving early warning of enemy moves to nullify surprise and reveal openings and opportunities for friendly initiatives. And finally, there was the *charge*, l'arme blanche, sabers raised, knee-to-knee, the impact weapon and shock action.

Aircraft, even in their crude and undeveloped state in the years before World War I, gave promise of becoming a far better horse. Certainly insofar as reconnaissance, interdiction, and the strategic role were concerned, the airplane bid fair to replace the horse. But the cavalrymen would have none of it. They didn't like machinery—they loved horses. As a minister for war in Britain once put it, to ask cavalrymen to give up their horses was like asking a concert violinist to give up his instrument and use the gramophone.

I remember an old Cavalry recruiting poster on the wall outside my office when I was teaching at West Point. It proclaimed: "The Horse is Man's Noblest Companion." That says it all. Logic indicated that the airplane should be assigned to the Cavalry, a combat arm with its already well-defined and extensive range of missions and doctrine. But the human factor, the mindset of the cavalrymen dictated another solution. So aircraft were assigned to the Signal Corps, a *service* not a combat arm. And for a whole generation Billy Mitchell and others struggled to break out of the "service" mold and secure for the airmen not only an organization appropriate for its full doctrinal potential, but also to secure resources sufficient to implement that potential.[53]

Holley brilliantly describes the conflict between instrumental values (the enormous combat potential of the airplane) and ceremonial values (Cavalry officers' hatred of the airplane and defaulting to place new technology in a science unit) in the Army as it confronted air power. His last statement even hints at the grammar and logic developmental responsibilities of military organizations. Holley says that air power lost and that the cavalry was ceremonially encapsulated, at least until armor permanently replaced the horse.

Here we see that institutional economics can shed new light even in military history. In most views of Air Force history, the Signal Corps to Air Service to Air Corps to Air Force is taken as a deterministic given—the way things needed to work out. Holley instead states that this deterministic evolution was not necessary and even a *mistake* caused by the inability of Army leadership to objectively assess the operational potential of the airplane because an elite of the Army—the Cavalry—refused to risk losing their horses! However, Holley also implies that a separate air organization, the ultimate aim of the air power zealots, was not the only goal. An independent air organization formed from a Cavalry air arm may also have been better than the historical Signal Corps air arm. Even though an independent space organization is not yet among us, could we have already fallen for the same mistakes that plagued air operations? Holley continues:

> All of us will probably agree that one of the most pressing problems confronting us as we escalate into the age of space is this: What organizational structure is best suited to the exploitation of space as an aspect of national defense? Should [Strategic Air Command], with its splendid track record of aggressiveness and exacting professionalism, have been the chosen instrument? Was a separate [Air Force] "Space Command" the best solution? Should such a command have taken over the research and acquisition functions for space from Systems Command, given the unusual character of the hardware? If a separate command is the approved solution, by the same logic, why not a separate "Space Force" entirely apart from the existing Air Force?...
>
> Has our organizational structure for space unwittingly fallen into the pattern that befell the airplane? Have we evolved our military space efforts as an ancillary *service* rather than as a combat arm? The language of those who speak knowledgeably on this subject and from positions of authority certainly reflects this perspective. We hear much of "mission support," an electronic bit stream providing the operating forces with pictures, words, weather reports, navigational signals, and the like, but only oblique and fleeting references to a combat role. As an under secretary of the Air Force put it: "The United States has never had weapons of any kind deployed in space and currently has no approved programs for the deployment of such systems in orbit." Of course, it is entirely possible that those in command may feel constrained by our current treaty obligations or by a sincere desire to avoid stimulating a politically undesirable arms race. They may feel constrained to avoid discussing space vehicles in a combat role, whether as "space superiority fighters" or as offensive strategic weapons. But surely the history of the early air arm and its organizational misadventure should give us pause. When it comes to national defense, which in the final analysis means *national survival,* treaties can be modified or abrogated by the prescribed procedure if need be. At the very least, with the message of our own institutional past ringing in our ears, it behooves us to study the organizational problem of space with the utmost care...
>
> If air arm doctrine at the end of World War I still defined the principal function of aircraft as observation, then logically it made sense to establish an Air Service in the years immediately following as an adjunct, subordinate to and supporting the combat arms. If we define our role in space as "mission support" for the operating forces, then will it not logically follow that the organization we build for space will be appropriate for a service or support role? Will we then have to wait

for some latter-day Billy Mitchell, some "space power" zealot, to buck the system and belatedly break out of the mold to develop a combat arm role for space?[54]

Institutional economists deny determinism in organizational development and that change has no predetermined direction, an assumption the General Theory of Space Power readily accepts. Therefore, society has the ability to apply *discretion*—to direct change toward predetermined goals.[55] Holley says that space organizational development to date has been towards evolving mission support roles (due to creating a support service Air Force Space Command) rather than a combat role (by denying space the Strategic Air Command combat model). Because there is no determinism in development, organizational choice is up to society to determine based upon the goals society wants from its military space organization.

Even though society has this choice, there will nonetheless be objectively better organizational choices than others. The objective metric with which to determine superior organizational methods is, again, maximizing space power by maximizing the development of space power logic and grammar through Paths 2 and 5. Broader ability (raw space power) is objectively better than lesser ability. How an actor uses space power is a separate issue unconnected to ability. A space power can use their space power to conquer or preserve even if they have the ability to do both. A space power without the ability to conquer will have no choice but to preserve, even under threat of annihilation. But there is also another question that remains unanswered. Which type of space organization will best be able to maximize space power given the particularities of the space environment: a combat-focused Space Force, a support oriented Space Command, a Coast Guard–like Space Guard, or some other type of organization? Holley asserts that this may not be a question answerable at this time, for we are missing a critical step:

> From the Air Service–Air Corps–Air Force perspective, there would seem to be two pressing organizational issues confronting all of us who think about the military in space. We must decide on the contours and dimensions of the space command or space force, whichever it turns out to be. But first we must develop our space doctrine because the doctrine we decide upon will inexorably influence the structure of the space organization we build...
>
> Doctrine, especially space doctrine, is vitally important. But we are confronted with the old chicken and egg dilemma: Which comes first? Doctrine will shape organization, but, until we perfect our organization for devising space doctrine, it is doubtful if we will be able to formulate a thoroughly satisfactory doctrine for space. The work of perfecting doctrine is complex; it calls for the willing and *informed* cooperation of many participants. Indeed, it calls for the exercise of substantial initiatives by participants in all the operating echelons. It cannot be left exclusively to a handful of specialists in a staff section. Consider, for a moment,

the very real differences between doctrine, on the one hand, and research and development— R&D—on the other. There are powerful economic incentives behind R&D. In our free, competitive, capitalist system, eager contractors are forever pressing technological innovations upon us. Their exciting proposals always outstrip our resources and force us to make hard choices. Nonetheless, the zeal of the contractors in coming forward with ever more remarkable developments virtually ensures an almost exponential technological progress.

But what economic motive force is there behind the formulation of doctrine? Where we pour literally billions of dollars into R&D, into ever more advanced hardware, we consign the task of generating space doctrine to scarcely more than a handful of staff officers already laden with a multitude of other tasks. And to make matters worse, the record of promotions for officers so assigned has not been such as to stimulate any great surge of eager talent into this exacting and demanding work. Clearly, in the absence of strong economic incentives to perfect our space doctrine, we would be well advised not only to concoct a highly efficient structure, an organization, but also appropriate *procedures* for devising sound doctrinal ideas. If we fail to do this *now*—in the immediate future—will we not be doomed to flounder ineffectually within the constraints of an organizational structure geared to a conception of the space mission long since outgrown?[56]

Although it seems clear that a dedicated organization for developing military space power (perhaps partnered with an organization dedicated to building space power in general) is the best way to promote military space power, it is likely we cannot fully ensure which type of organization (force, corps, guard, etc.) would be best at accomplishing this mission. What we do know is that, since development isn't deterministic, whatever decision on the culture of the service we make will drag military space power (and perhaps even space power in total) in some direction, positive or negative. Every decision to change military space organization, or decision to let the status quo remain, will indelibly leave its stamp on space power development. Because this is such an important revelation, we must attempt to understand space power as soon as possible in order to inform correct decisionmaking, even when not deciding to act is an action in itself.

Since space power is developed in specific ways (the five paths) that are in no way connected to air power development and that the "tooth to tail" metric (a losing one) should rather be "head to tail" (a winning one), the general theory indicates that an independent military space organization is to be supported over the current air and space force model. However, the type (force, corps, guard, or other) is not yet settled (though it should be noted that any of them would undoubtedly be better than the status quo).

The model of the independent military space organization will be determined by how well it is projected to foster space power development. How best will each model promote Paths 2 and 5 development in the Logic and

Grammar of Space Power? It should be remembered that policy does have a role to play in the decision (do we want a combat-centric warfighting type or a development-oriented guardian-type space military?). We must also look at the space power environment itself. The six space environment characteristics in the space power hierarchy of access are relevant. What type of service will best be aligned to exploit all of the characteristics?

The only way we can gain assurance of our decision is to take Holley's advice. We must first form the prerequisite organization to develop doctrine for space in an objective a way as possible. It is likely we will need to form the organization against Holley's advice and begin it as a small collection of staff officers (however, this time dedicated primarily to academic pursuit, such as the faculty of the historic Naval War College and Air Corps Tactical School in their intellectual primes) and leave them to connect to the operating forces over time and as interested combat force volunteers become active in the project to develop military space doctrine. Only then, where force, corps, and guard activists can meet in intellectual battle, judged by objective (if not disinterested) officers will we find the best decision on what type of military space organization is best.

General Theory Advice on Vision

The General Theory of Space Power cannot advocate the ultimate ends required of space power. However, it does suggest that organizations dedicated to space power development must focus on increasing their ability to operate in space for all purposes, not simply one type of applied space power. NASA, so far, has organizationally focused on the political logic of space power to the detriment of increasing general ability to operate in space. Likewise, the military space commands have, since the Eisenhower administration, stripped them of a general space development mandate and have focused exclusively on the military logic of space power. Both organizations have neglected developing the common ability to operate in space that the General Theory suggests is the true fount of space power, and this neglect has manifested itself in an aimless and disappointing American space program, even though American space power is still the most mature in the world.

However, there is clear evidence that the military had the most expansive visions of space power in the early Space Age. This expansive vision was the natural product of the military mind that, opposite of the common stereotype of being dull and inflexible, was innovative and holistic from its long history of service to both American exploration and development, as well as defense.

By applying the lessons of the General Theory of Space Power to develop a new military space organization with a renewed mandate to promote the development of American space power in general, the United States will dramatically advance its effectiveness and efficiency in developing the Logic of Space Power.

But developing enhanced space power logic capabilities is only half of the problem. What policies must the military space organization enact in order to develop the skills, technology, and machinery necessary to fully exploit the space environment? The next chapter will discuss how to organize for effective development of the Grammar of Space Power.

Chapter 3

Organizing for Effective Development—Grammar

True space power is built in the Grammar Delta. Logic Delta operations develop a nation's skill at translating new technology into instruments of power, but activity at the Grammar Delta builds the tools that serve as the foundations of space power. Optimizing the efficiency of Grammar Delta operations offer the possibility of dramatic leaps in the nation's access to the space environment and promise rapid technological and economic growth. This chapter delves deeper into space power grammar development and confront its nature, challenges, and possibilities in order to better grasp how we might be able to improve our capability to master Grammar Delta operations.

Space Power and Technology Development Strategy

Space power is limited today primarily due to technological reasons— we simply do not have the requisite level of technology required to fully exploit the continuum of advantages the space environment offers. Whereas in a mature future space power decisions will be based on which planet to colonize or which star system to explore (resource-based restrictions due to lack of sufficient elements—i.e., starships—to do everything we would like), right now the limiting factor is one of access: we don't have a sufficient level of technology to colonize anywhere with any relevant number of people and equipment (our existing elements are incapable of doing anything at any scope). To develop elements (production, shipping, and colonies) capable of exploiting the higher levels of the space power hierarchy (resources and new worlds), advances in technology—Path 1 development—will play a central and perhaps dominant role in space power development for the indefinite future.

Because of technological development's supreme importance to space power development, it is necessary to understand how technological improvements come to pass so that we may utilize that understanding to improve their efforts to develop space power. In their book *The Strategy of Technology*, Stefan Possony and science fiction legend Jeffrey Pournelle offer a model of the technological development process (called the Technological Process) which is highly instructive. They offered this model to inform the commanders of what they considered the "Technological War," the Cold War battle of technology waged between the United States and the Soviet Union. Even though space power is not currently the battlefield between two or more major hostile powers (perceived tensions between the United States and China notwithstanding), the Technological Process model is highly valuable to space power architects.

The Technological Process model envisions four distinct phases that any technology goes through from a basic scientific idea to a fielded operational system. These phases are the Intellectual Breakthrough phase, the Invention Breakthrough phase, the Management Breakthrough phase, and finally the Engineering Breakthrough phase.[1] Possony and Pournelle admit that the four stages of the Technological Process are merely indicative of "broad areas of human activity" and that the four-step division is an illustrative rather than concrete delineation.[2] Regardless, through these four phases, any Path 1 technology development program can be studied and explored.

The Intellectual Breakthrough phase begins the Technological Process. In this phase, scientists (Possony and Pournelle call these people "men of genius") make discoveries and propose theories which fundamentally overturn accepted scientific understanding. Such discoveries are called breakthroughs because they *eliminate restrictions imposed on scientific thought* imposed by classical science up to the point of discovery.[3] By unleashing new understanding of basic principles, the Intellectual Breakthrough phase opens new vistas of thought that increases the conceivable possibilities over many different vistas of application. As such, the Intellectual Breakthrough establishes the basic science foundation of new applications that may be built using this new discovery or fundamental change in understanding.

Possony and Pournelle note that intellectual breakthroughs only lay the foundation for future applications. The Intellectual Breakthrough phase does not increase any capabilities and is merely the necessary precondition for a new capability. Two important characteristics of this phase merit mention. Firstly, intellectual breakthroughs occur unpredictably and cannot be anticipated in advance. Put bluntly, genius cannot be set to a timetable. This is a critical point with large ramifications that will be discussed later. Secondly, it normally takes a great deal of time for the implications of a new scientific breakthrough to become fully appreciated (the authors' opinion is two generations' time) and the breakthrough is potentially able to proceed to the next phase. The Intellectual Breakthrough phase is unpredictable and difficult (if not impossible) to manage. However, it is among the most valuable events that can possibly happen, for once an intellectual breakthrough occurs, adapting the discovery to improve human life and capability becomes possible. The Intellectual Breakthrough phase translated into the General Theory increases the knowledge portion of the foundation of space power at the base of the Grammar Delta. Beyond this, the Intellectual Breakthrough phase does not interface with the Grammar or Logic Deltas themselves in any way. Intellectual breakthroughs help lay the foundations of space power—they are not space power themselves.

Once the intellectual breakthrough becomes sufficiently understood, the process of application can begin. The second phase of technological development, the Invention Breakthrough phase, ends the basic science phase begins this application process. This second step in the Technological Process is devoted to translating the new understanding of basic science developed in the Intellectual Breakthrough phase into a device with some useful purpose. Key to this step is the "instinctive or intuitive confidence that something should work and the first rough test of whether [this new technology] will in fact work."[4] This phase is firmly in the realm of technology rather than science (though it is still very much a creative art), but new advances in science may take place here as new science is forged into technological innovations. The Invention Phase can be considered as the "applied science" phase. It may also take many years to complete, though it is generally completed far faster than the Intellectual Breakthrough phase. The end product of the Invention Phase can be a proof of concept demonstration or a breadboard model of a new product. It is an important step, but new space power elements are not yet developed and we have yet to build new space power. We are simply one more step through the Technological Process.

The third step of the Technological Process takes place outside of the purview of the scientist (step 1) and the applied scientist or design engineer (Phase 2) and into the hands of the technical manager. In this Management Breakthrough phase, managers from industry or the military recognize that an invention from Phase 2 has potential importance and value, and decide to allocate resources to translate the invention or demonstration into a materially useful product.[5] The Management Breakthrough phase has major implications on future capabilities because it is in this phase that the potential power from basic and applied science is chosen to begin development of new elements of space power. This phase normally ends with a decision to place an invention into real production. It is in this phase that we emerge from the foundation of the Grammar Delta into the base of the delta itself.

The purpose of this phase is to gain an advantage in time or strength over competitors (the logic of power), which economically can mean market advantage and militarily would produce a strategic advantage over potential adversaries. In short, the General Theory states that this phase is the decision point to initiate the development of space power. Because the Logic of Space Power is relative (i.e., that space power strategy relies on the decisions of other actors and potential adversaries) the Management Breakthrough is not a sterile static decision, but must be dynamic in relation to the moves of other players in the space environment.

However, the Management Breakthrough phase is critical for space power development (the Grammar of Space Power) because it is in this phase where a vision of space power can substantially change which space power elements are developed. Shall a space power develop heavy nuclear launch rockets, orbital stations, human spaceflight technology, or advanced microelectronics for small satellites? It is in the Management Breakthrough phase that space power "art" through visions of space power (explored in Chapter 2) can substantively alter space power development.

Although critical questions of the link between the logic and grammar of power (see Chapter 1) must be addressed in the Management Phase, it is a relatively quick phase in relation to the other stages of the Technological Process. It is also the first step in translating a piece of the foundation of space power into space power itself. In the final phase of the Technological Process, a new element of space power is finally produced.

The fourth, and last, phase of the Technological Process is the Engineering Breakthrough. In this last phase, the "invention chosen by management is developed as a system and produced in quantity."[6] Critical to this step is the development of a prototype element of the system itself, the bridge between the breadboard invention and a full scale production model. This phase, according to Possony and Pournelle, is completely immersed in the realm of technology and engineering, meaning that in the General Theory it leaves science (and the foundation of the Grammar Delta) behind and becomes firmly an activity of the base of the delta. The prototype is a completely new space power element and can be combined with other elements to increase space power access. Success in the Engineering Breakthrough phase is the *only* Technological Process step in which space power access and ability is developed for national use. At the end of this phase, a new space power element (or elements) is placed in the Grammar Delta for exploitation by the space power. At this point, the Technological Process can offer no more to the development of space power in this specific type of element (although incremental improvements and more advanced method of producing the same effects will always be useful). Path 1 development of new space power elements, is complete.

It is important to note that the Engineering Breakthrough phase as the culmination of Path 1 development does not necessarily require large scale fielding to be complete. One workable, reproducible prototype is all that is necessary for the grammar portion of Path 1 space power development to be complete. Decisions on the number of complete units required to be effective is wholly a question of space power logic, and the answer may change based on outside conditions. However, grammar is strictly a method of building and enhancing the toolkit available to the Logic Delta. Once an element is available for production at whatever level desirable, the Technological Process has been successfully concluded and space power development then moves on to finding new ways to combine this element with others to enhance space power access.

Disruptive Space Power Development

Due to the nature of discovery and the Technological Process, space power does not develop smoothly. Being a type of economic development, it relies on new combinations of elements to expand the ability of space power to accomplish things in space. Schumpeter describes this discontinuity in development:

> Why is it that economic development in our sense does not proceed evenly as a tree grows, but as it were jerkily; why does it display those characteristic ups and downs?...
>
> The answer cannot be short and precise enough: exclusively *because the new combinations are not, as one would expect, according to the general rules of probability, evenly distributed through time*—in such a way that equal intervals of time could be chosen, in each of which the carrying out of one new combination would fall—but *appear, if at all, discontinuously in groups or swarms.*[7]

Technological innovation, Schumpeter's new combinations, cannot be counted to arrive on a regular schedule. Indeed, many innovations come in groups. The classic example is the steam engine. Although the steam engine in itself was a single innovative application of thermodynamics and mechanics to develop a new type of engine, the resulting economic impact of the technology (steam engine) and innovations resulting from the technology was nearly incalculable. The steam engine allowed the railroad to revolutionize ground transportation, eliminated the need for sails on sea craft, brought widespread mechanization of factories and industrial production, and laid the foundation for rapid economic expansion in almost all avenues of human activity. However, once the steam engine was fully integrated into the economy, the expansion could not be sustained indefinitely. The steam engine, while powerful, could propel human industry only so far, and once fully utilized, it could add no more speed to the economy. The economy would need to wait for another technology—gas engines or the computer—to expand the economy in great leaps again.

Space power develops in similar fashion. The development of high-quality liquid rockets allowed humans to scrape space. Satellites would have been a mere frivolity without a long-duration power system allowed by regenerative batteries and photovoltaic cell technology. Many space enthusiasts now say that we have reached the peak of what we can do with the liquid chemical rocket (though many also disagree) and for space power to be significantly enlarged again we would need a much more powerful engine, such as nuclear thermal rockets. Whether this is true or not, few could argue that an engine that could place a payload into orbit for $10 per pound would lead to a rapid expansion of space activity over all applications—projects that would be useful but cost ineffective at $10,000 per pound to orbit. Indeed, such a new cheap engine could be termed a space power *revolution.*

Since space power development is discontinuous as new technologies are applied to the elements and new concepts are used to combine them into new avenues of access, ultimately expanding ability, this discontinuous development will often result in space power revolutions. A space power revolution can be defined as where the elements change so dramatically we can say that the entire Grammar Delta is shifted (although space power logic always remains the same) into a completely new expression of space power, often becoming a game changing event in the current balance of space power. Examples of a revolution from the sea power realm would be the invention of sails, steam power, or carrier-based aviation.

Space power revolutions will likely take place from one of two major avenues of technology: having the main thrust of current space power exploitation climbing the hierarchy of space power environmental characteristics (for instance, from our current "orbital mechanics" phase to "solar energy" through the construction of large-scale solar power satellites for Earth-bound power consumption), or developing a technology that expands access by an order of magnitude (i.e., opening the Moon to mass travel from low Earth orbit, Mars, inner solar system, entire solar system, interstellar flight).

Note that both avenues are Path 1 development approaches: dramatic changes in the effectiveness of space power elements. Revolutions, then, occur at the element level. However, revolutions do not always occur simply due to new technology.

Technical versus Economic Logic

Another key insight into space power development is to identify a very dangerous fallacy arising from the nature that space activity has with high technology—the fetishization of technology itself over its utility for space power purposes. There is a fundamental difference between technological sophistication and economic utility, even though both developing the technological sophistication and determining which technological processes will provide the highest profit are both vitally important. This fundamental truth has often been ignored in the space community. Schumpeter explains:

> [The economic problem] must be distinguished from the purely technological problem of production. There is a contrast between them which we frequently witness in economic life in the personal opposition between the technical and commercial manager of an enterprise. We often see changes in the productive process recommended on one side and rejected on another; for example, the engineer may recommend a new process which the commercial head rejects with the argument that it will not pay. The engineer and the businessman can both express their point of view thus: that their aim is to run the business suitably and that their judgment derives from the knowledge of this suitability. Apart from misunderstandings, lack of knowledge of facts, and so forth, the difference in judgment can only come from the fact that each has a different kind of appropriateness in view. What the businessman means when he speaks of appropriateness is clear. He means commercial advantage, and we can express his view thus: the resources which the provision of the machine would require could be employed elsewhere with greater advantage. The commercial director means that in a non-exchange economy the satisfaction of wants would not be increased, but on the contrary reduced, by such an alteration in the productive process. If that is true, what meaning can the technologist's standpoint have, what kind of appropriateness has he in mind? If the satisfaction of wants is the only end of all production, then there is indeed no economic sense in having recourse to a measure which impairs it. The business leader is right in not following the engineer, provided his objection is objectively correct. We disregard the half-artistic joy in technically perfecting the productive apparatus. Actually, in practical life we observe that the technical element must submit when it collides with the economic. But that does not argue against its independent existence and significance, and the sound sense in the engineer's standpoint. For, although the economic purpose governs the technical methods as used in practice, there is good sense in making the inner logic of the methods clear without practical barriers.

This is best seen in an example. Suppose a steam engine in all its component parts complies with economic appropriateness. In the light of this appropriateness it is made the most of. There would be then no sense in turning it to greater account in practice by heating it more, by letting more experienced men work it, and by improving it, if this would not pay, that is if it could be foreseen that the fuel, cleverer people, improvements, and increase in raw materials would cost more than they would yield. But there is good sense in considering the conditions under which the engine could do more, and how much more, which improvements are possible with present knowledge, and so forth. For then all these measures will be worked out ready for the time when they become advantageous. And it is also useful to be constantly putting the ideal beside the actual so that the possibilities are passed by, not out of ignorance but on well-considered economic grounds. In short, every method of production in use at a given time bows to economic appropriateness. These methods consist of ideas not only of economic but also of physical content. The latter have their problems and a logic of their own, and consistently to think these through—first of all without considering the economic, and finally the decisive factor—is the purport of technology; and in so far as the economic element does not decree otherwise, to put them into practical effect is to produce in the technological sense...

But the economic and the technological combinations, the former concerned with existing needs and means, the latter with the basic idea of the methods, do not coincide. The objective of technological production is indeed determined by the economic system; technology only develops productive methods for goods demanded. Economic reality does not necessarily carry out the methods to their logical conclusion and with technological completeness, but subordinates the execution to economic points of view. The technological ideal, which takes no account of economic conditions, is modified. Economic logic prevails over the technological. And in consequence we see all around us in real life faulty ropes instead of steel hawsers, defective draught animals instead of show breeds, the most primitive hand labor instead of perfect machines, a clumsy money economy instead of cheque circulation, and so forth. The economic best and the technologically perfect need not, yet very often do, diverge, not only because of ignorance and indolence but because methods which are technologically inferior may still best fit the given economic conditions.[8]

Technological efficiency and economic efficiency are not the same, even in space power development where technology is critical to expanding access and ability. However, this difference is not often respected in space organizations. NASA has been criticized in the past for building "space Ferrari" rockets where "space Chevy" rockets would do. The obsession with technological elegance over economic reality is well showcased by NASA's decision to cancel the technologically inelegant but flying DC-X and DC-XA experimental rocket in favor of the bleeding edge X-33 single-stage-to-orbit rocket which required significant advances in engines, cryogenics, and materials in order to work. As Schumpeter says, economic logic trumps technical logic in sustainable development. Needless to say, the DC-X flew many times while the X-33 was a very expensive development project that never resulted in flight hardware.

Space development officials in both private and public life must respect the difference between economic and technological logic in order to develop space successfully. There is a proper role for research and development, as Schumpeter explains above, but function must trump form in order to accomplish real objectives. However, technology can perform different functions.

Influence versus Access Technology

Recall that the General Theory's definition of space power is the ability to do something in space. This definition can be broken down into two parts that imply two separate types of technology: "do something" and "in space." "Do something" elements are examples of Influence Technology, technology designed to operate in space for some useful purpose that generates economic, military, or diplomatic power. Examples of Influence Technology would be communications relays, high definition cameras, or solar power cells that can perform useful work while operating in space. These elements are classified as Influence Technology because the Logic of Space Power is to influence an adversary or competitor to change his ways to become more acceptable to the agent using space power. Influence Technology is technology that can be used in space to generate power (influence) from space.

The second critical piece of our definition, "in space," requires Access Technology. Essential to developing the Logic Delta's ability is to build the Grammar Delta's access. Access, again, is the capacity for an agent to place something in a specific area of space. Access Technology is that technology which permits progressively larger sections of space to be open to exploitation. Access Technologies are generally engine technology and technology to protect systems from the space environment. Examples would be chemical or electrical propulsion systems as well as the radiation shielding necessary for elements to survive in particularly harsh environments such as the Van Allen belts or the Jovian moons. Access Technology's only use is to enable Influence Technology to accomplish its mission.

It is somewhat paradoxical, then, that in space power development Access Technology is generally more important than Influence Technology. The primary reason that Access Technology is more important is because access is most often the "long pole in the tent" for most things we can envision doing in space. We often have the equipment necessary to do something useful in space (or can easily modify terrestrial equipment to work in space) but have no physical or economical way to transport it into an advantageous part of space and maintain the equipment while there. Exploitation is relatively easy. Access is hard.

A second reason Access Technology is more important is that Access Technology is often space-specific in ways Influence Technology is not. Space engine development will often only be pursued by space agents where Influence Technology that can be used or easily adapted for use in space will often be developed by other industries and operations. Therefore, space agents will often need to shoulder the entire fiscal and operational burden for opening increasingly larger areas of space for development in ways that will simply not be necessary for most pieces of Influence tech. Due to the uneven burden of Access Technology versus Influence Technology, Access Technology should be considered more important and taken more seriously by space power development agents.

A third reason for the importance of Access Technology is that once a new area of space can be exploited, we have the ability to use all previous pieces of Influence Technology in that new area. This dramatic increase in ability is much larger than that engendered by the development of a single new piece of Influence tech. Ability skyrockets with an increase in access in what is likely a geometrical relationship where the ability increase from Influence Technology is probably closer to a linear relationship.

Both Influence Technology and Access Technology are critical to space power development. Influence Technology allows for power to be produced from space and is critical to the Logic of Space Power. Access Technology, alternatively, is critical to the Grammar of Space Power because it generates new access to the space environment. Even though both are important, Access Technology is the sole province of space power agents and must be considered on a theoretical and practical level the most important types of space technology to foster the expansion of space power. Harnessing the power to deliberately develop technologies and deriving real power and advantage from its development is the focus of the next section.

The Technological Warfare Campaign

Technological warfare is a subject many have heard of but few truly understand. Colonel John M. Collins says "Technological warfare, which connects science with strategies, operational art, and tactics, endeavors to make rival armed forces uncompetitive, preferably obsolete.[9] Possony, Pournelle, and Kane, in their second edition of *The Strategy of Technology*, offer a broader and, perhaps, more correct view by defining technological war as

> the direct and purposeful application of the national technological base and of specific advances generated by that base to attain strategic and tactical objectives. It is employed in concert with other forms of national power. The aims of this kind of warfare, as of all forms of warfare, are to enforce the national will on enemy powers; to cause them to modify their goals, strategies, tactics, and operations; to attain a position of security or dominance which assists or supports other forms of conflict techniques; to promote and capitalize on advances in technology to reach superior military power; to prevent open warfare; and to allow the arts of peace to flourish in order to satisfy the constructive objectives of society.[10]

This book intends to explore and model a subset of technological war: a single technological breakthrough and its potential national power implications. It will model this single "technological campaign" using widely accepted graphical models from both technology and innovation theory as well as a model of Clausewitzian warfare. After developing the technological warfare model, we will close with some insights from the model that may be of use to technological warriors in the future. In order to begin, we must first review the importance of technological strategy.

The Strategy of Technology

Possony, Pournelle, and Kane's book *The Strategy of Technology* is a foundational work for understanding technological warfare. In the book, they claim that like any type of warfare, a belligerent in a technological war requires a strategy in order to be effective. They state the requirements as:

> A technological strategy would involve the setting of national goals and objectives by political leaders; it would be integrated with other aspects of our national strategy, both military and nonmilitary (Initiative, Objective, and Unity of Command); it would include a broad plan for conducting the Technological War that provided for surprising the enemy, pursuing our advantage (Pursuit), guarding against being surprised (Security), allocating resources effectively (Economy of Force), setting milestones and building the technological base (Objective), and so forth.[11]

These requirements are familiar to any student of strategy, combining ways and means to achieve specified ends and designed with a healthy (but not dogmatic) respect for the principles of war. However, technological warfare is not a carbon copy of classical warfare. While technological warfare shares many of the goals of classical warfare, it is fought differently due to the differing natures of technology and combat forces. Thus, a technological strategy, and a model of technological warfare, requires a model of technology to anchor its foundation to the nature of its subject matter.

THE NATURE OF TECHNOLOGY

A common model used to describe the nature of technology and technology advances is the S-curve (Figure 3.1). S-curves were the preferred method by which Possony et al. developed their theory of technological warfare:

> One of the most easily observed phenomena of technology is that it moves by "S" curves, as illustrated in Figure One [reproduced in this book]. Take for example speed; for centuries the speed of military operations increased only slightly as each side developed better horses. Then came the internal combustion engine. Speed rose sharply for a while. Eventually, though, it flattened out again, and each increase took longer and longer to achieve.
>
> To illustrate the S-curve concept, consider the development of aircraft, and in particular their speed. For many years after the Wright brothers, aircraft speeds crawled slowly forward. In 1940, they were still quite slow. Suddenly, each airplane designed was faster, until the limits of subsonic flight were reached. At that point, we were on a new S-curve. Again, the effort to reach transsonic flight consumed many resources and much time, but then the breakthrough was made. In a short time, aircraft were traveling at multiples of the speed of sound, at speeds nearly two orders of magnitude greater than those achieved shortly before World War II.[12]
>
> Note that the top of one S-curve may—in fact usually will—be the base of another following it. Although the stream moves on inexorably, it is possible to exploit one or another aspect of technology at will. Which aspect to exploit will depend on several factors, the most important being your goals and your position on the S-curve.[13]

Perhaps the most extensive exploration of S-curves is Richard N. Foster's book *Innovation: The Attacker's Advantage.* In it, he succinctly explains the S-curve and its shape:

> The S-curve is a graph of the relationship between the effort put into improving a product or process and the results one gets back for that investment. It's called the S-Curve because when the results are plotted, what usually appears is a sinuous line shaped like an S, but pulled to the right at the top and pulled to the left at the bottom.
>
> Initially, as funds are put into developing a new product or process, progress is very slow. Then all hell breaks loose as the key knowledge necessary to make advances is put into place. Finally, as more dollars are put into the development of a product or process, it becomes more and more difficult and expensive to make technological progress. Ships don't sail much faster, cash registers don't work much

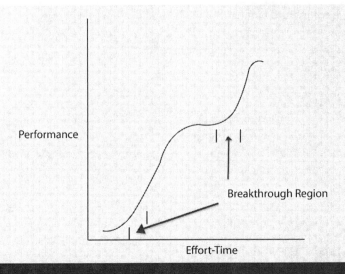

Figure 3.1 Technology S-Curve

better, and clothes don't get much cleaner. And that is because of limits at the top of the S-curve.[14]

Although there are a number of interpretations of the horizontal and vertical variables of the S-curve, the most popular usage is to label the dependent vertical variable "performance" of the technology in terms of user demands (for instance, horsepower on engines or top speed of aircraft). The horizontal independent component is in terms of "effort" toward improving the technology's performance. The most common measure is "lab hours" or "man hours" which is time multiplied by the number of people working on the problem. Thus, the independent variable is not simply time, but time multiplied by effort to create an adjusted time able to be shortened or lengthened depending on the actions of the agent conducting the research.

The shape of the S-curve is vitally important to prosecuting technological warfare and devising a technological strategy. However, contra Possony et al., S-curves alone are not enough to adequately model technological warfare or even a technological war "campaign," a single military-relevant breakthrough. In order to adequately model the technological war campaign, we must enter the realm of military strategy to bridge the gap between technology and warfare.

THE PRINCIPLE OF PURSUIT

Fortunately, Possony et al. provide the connection with which to bridge the gap through the idea of technological pursuit. They state:

> Whether the breakthrough is a surprise to the enemy or is an advance that he anticipates but cannot counter, the side making the breakthrough should plan for technological pursuit to maximize the gain made possible by the new advantage. Pursuit has proved difficult in warfare. The losses sustained in winning the battle frequently have reduced the momentum of the winner. Also, uncertainty about the conditions of the loser has made the winner act with caution.
>
> In technological conflict pursuit is facilitated by the circumstances surrounding the breakthrough. Rather than causing losses, the technological success increases the power of the side making the advance, and success often heightens morale. The breakthrough can reduce the amount of uncertainty about the enemy's technology position.
>
> These circumstances point out clearly that significant technical advances must be exploited. The concept of pursuit has a valid role in technological conflict.[15]

The principle of pursuit (also called the principle of continuity) comes from Clausewitz. Michael Handel says "This principle states that commanders must exploit an advantage by keeping the enemy under unrelenting pressure, thereby denying him respite or time to regain his equilibrium. The underlying logic is universal: it makes no sense for the side that has gained an advantage to give an opponent the chance to renew his resistance later on."[16] In Clausewitz's words:

> Once a major victory is achieved there must be no talk of rest, of a breathing space, of reviewing the position or consolidating and so forth, but only of the pursuit, going for the enemy again if necessary, seizing his capital, attacking his reserves and anything else that might give his country aid and comfort....
>
> All that theory requires is that so long as the aim is the enemy's defeat, the attack must not be interrupted. If the general relinquishes this aim because his considers the attendant risk too great, he will be right to break off.... Theory would blame him only if he does so in order to facilitate the defeat of the enemy.[17]

Thus, the principle of pursuit advocates attack for maximum gain as the proper response to a victory. However, Clausewitz points out that the risk attendant to a continuing attack will eventually outweigh the potential gains. Once this happens, we have reached the culminating point of the attack. Returning to Clausewitz:

> Success in attack results from the availability of superior strength, including of course both physical and moral.... The attacker is purchasing advantages that may become valuable at the peace table, but he must pay for them on the spot with his fighting forces. If the superior strength of the attack—which diminishes day by day—leads to peace, the object will have been attained. There are strategic attacks

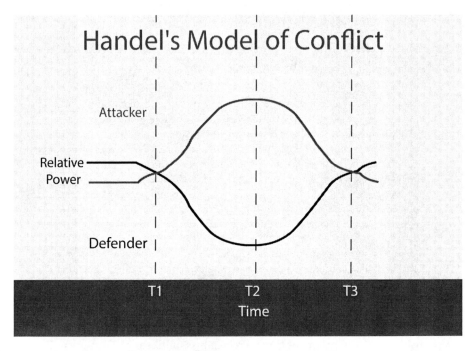

Figure 3.2 Handel's Model of Clausewitz's Culminating Point of Attack/Victory (Handel, p. 186).

that lead up to the point where their remaining strength is just enough to maintain a defense and wait for peace. Beyond that point the scale turns and the reaction follows with a force that is much stronger than that of the original attack. This is what I mean by the culminating point of the attack.[18]

Once the culmination point has been reached, the principle of pursuit is no longer valid and the attacker no longer has any remaining increase in relative power to the defender from his original victory.

Handel offers a useful visual depiction of Clausewitz's thoughts concerning victory, pursuit, and the culmination point of the attack, reproduced in Figure 3.2. Handel describes the model:

> The attacker (red) and defender (blue) have roughly comparable strength when the war begins at T1. The attacker achieves strategic and operational surprise and rapidly advances. As the attacker continues to advance (from T1 to T2), he gathers strength as the defender weakens. As long as he can both advance and gain strength, [the attacking general's] decision should be dictated by the principle of continuity [pursuit]. Gradually the defender regains his equilibrium while the attacker loses much of his momentum (beginning at T2)—his forces are getting further and further away from their bases of supply; his lines of communications are extended, and his flanks are exposed. At the same time the defender continues to fall back on

his own supply bases; his communication lines become shorter, and the population friendly. Time works in his favor. At T2, on the curve the attacker has reached the *peak of his power* relative to the defender, but as he begins to grow *relatively weaker* (from T2 onward), the defender is becoming *stronger*. At T3 on the curve, the defender overtakes the attacker and the momentum of a counter-attack is on his side.[19]

With this simple visual model, Handel distills Clausewitz's insight on both the principle of pursuit and the culminating point of the attack. With this graphical representation of conflict from military theory, we can now relate the nature of technology to the nature of warfare to produce a graphical model of a technological war campaign.

A Technological War Campaign Model

For our model, the S-curve is adjusted only slightly from its original intent. Instead of the S-curve only modeling technological performance, we adapt an adage from I.B. Holley to develop the concept of military perform-ance:

> Superiority in weapons stems not only from a selection of the best ideas from advancing technology but also from a system which relates the ideas selected with a doctrine or concept of their tactical or strategic application, which is to say the accepted concept of the mission to be performed by any given weapon.... New weapons when not accompanied by correspondingly new adjustments in doctrine are just so many external accretions on the body of an army.[20]

Therefore, we will define military performance as a combination of techno-logical performance of a given new technology, doctrinal maturity of the new technology, and progress completed towards full deployment of the technol-ogy for military use. Now the model can be completed.

The Technological Warfare Campaign model is shown in Figure 3.3. A technological warfare campaign is an event where a single militarily relevant technology breakthrough occurs and aggressor (the belligerent that develops the innovation) strategists decide to capitalize on the breakthrough by exploit-ing the temporary increase in power against the conservator (the belligerent which will be influenced by this breakthrough for aggressor purposes). The model itself is the combination of an S-curve model (above) and Handel's model of conflict (below) linked by a common independent variable axis Effort-Time, which can be measured as man-hours of effort assigned to devel-oping the new weapon's technological performance, maturing the new weapon's military doctrine for use, and completing a full deployment of the new technology to field units. By linking these graphs through their common

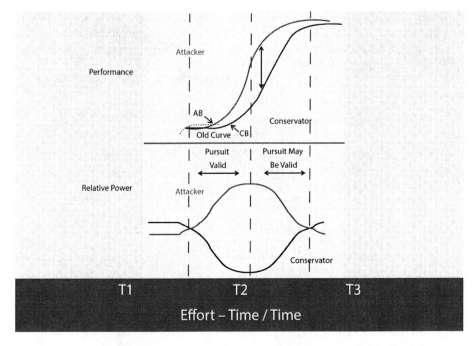

Figure 3.3 The Technological Warfare Campaign Model

variable, both graphs can be used simultaneously to analyze the campaign. But both graphs must be slightly altered to ensure compatibility.

The S-curve in the model is adjusted by altering the dependent variable (Y) axis from being solely a measure of technological performance to military performance to accommodate both technological and doctrinal development in addition to fielding the technology. This alteration is essential because new technology is meaningless to the military unless it also has copies of the technology to use as well as some idea of how to use the technology effectively. The S-curve is also adjusted by the addition of a new S-curve to account for the conservator's adoption of the aggressor's technology or a different new technology as a countermeasure to the aggressor in order to regain its lost relative power due to the aggressor's breakthrough and bring the technological war campaign to a close.

Alternatively, the Handel chart below on the model must be altered in its independent variable (X) axis in order to take it away from simple time to "effort-time" to match with the S-curve concept. This is logical since the time it takes for either an aggressor or conservator to adopt a technology is dependent on the number of scientists, engineers and strategists assigned to learn how to exploit this new weapon.

The technological war campaign begins at T1. At this time the aggressor achieves a military breakthrough at point AB. This is the point where a new technology achieves performance parity with the old technology. Note this is not the technological breakthrough point which starts an S-curve onto a rapidly increasing slope familiar to most S-curves. The technological breakthrough happens before parity, earlier than T1. At point AB, the aggressor starts fielding new weapons available from the breakthrough and from T1 on the result of fielding the advanced weapons, the relative power increases for the aggressor and decreases for the conservator. From T1 to T2, the aggressor is gaining in relative power over the aggressor. However, at some point between T1 and T2 the conservator reaches a military breakthrough (point CB) where he can begin to field effective countermeasures to the aggressor's technology. Beyond point CB, the conservator begins to arrest the aggressor's increasing advantage. This arrest is due primarily to the nature of S-curves. The aggressor will eventually reach a time when his S-curve begins to flatten out as the curve's slope decreases. Conversely, the conservator's S-curve will be relatively younger and at a higher slope as the same point in time. Essentially, the aggressor is teaching his maximum military potential from the technology while the conservator has room to grow. At T2, the largest vertical difference in military performance between the aggressor and the conservator, the aggressor is at the highest level of relative advantage in the campaign, but the conservator has also completely arrested his ascent. From T2 to T3, the conservator is still relatively weak compared to the aggressor, but the tide has turned. The aggressor no longer has such a commanding lead and cannot as effectively demand concessions from the conservator due to the efforts of the campaign, for the conservator knows he will soon negate the aggressor's advantage in this technology. At T3, both the attacker and conservator have fully matured and deployed the technology that started the campaign and its countermeasure. Therefore, the technology can offer no more relative power to either belligerent and the campaign is concluded. The technological war must continue through the development of other technologies.

Note that when the technological warfare campaign is over at T3, this does not mean that the relative power between the aggressor and the conservator is back into parity. On the contrary, at the end of the campaign the aggressor, if successful, has achieved the objectives of the campaign and has gained a permanent advantage over the conservator. The return to parity of relative power at T3 is simply the relative power the military breakthrough that started the campaign is currently producing for the aggressor. It will eventually lose its utility even if it has earned the aggressor large concessions in other matters. For instance, an aggressor fielding a superior attack submarine may allow a country to win the rights to a disputed island from a conservator diplomatically if the conservator knows he cannot mount an effective military resistance. Even when the conservator is able to counter the aggressor's submarine, the loss of the disputed islands is a loss that extends far beyond the technological war campaign.

Victory in technological war, according to Possony et al., "is achieved when a participant has a technological lead so far advanced that his opponent cannot overcome it until after the leader has converted his technology into decisive weapons systems."[21] This is true in technological war, but a technological campaign cannot be expected to win the war by itself. Therefore, "victory" in the sense of a technological war campaign is simply to achieve the campaign's objective (i.e., what the aggressor wants to gain from this temporary increase in relative power—relative power increase in itself should not be a goal by definition of technological war above) before the advantage of the new technology is negated either by conservator adoption of the technology or a successful countermeasure. The technological campaign must be fought for limited objectives. By understanding the implications of each component of the campaign model we can develop some insight into which objectives are achievable and when the aggressor should make decisions regarding follow-on campaigns.

Pursuit and the Culminating Point: Insight from Clausewitz

The primary insight that can be gained from Handel's model applied to technological war is to help determine the culmination point of the campaign and assess if the target objective of the campaign can be reached. We must always remember that what matters to the aggressor in a technological war campaign is not the relative strength of the opponent, but the achievement of the campaign's object.[22] Therefore, we must use Handel's portion of the model to assess where the object may be achieved relative to the time span of the campaign and whether this time appears favorable to the aggressor.

Even if the campaign is successful and the aggressor reaches his objective before the campaign's culmination point, Clausewitz offers more advice to the technological strategist:

> Lastly, even the ultimate outcome of a war is not always to be regarded as final. The defeated state often considers the outcome merely as a transitory evil, for which a remedy may still be found in political conditions at some later date.[23]

Technological strategy, like strategy itself, never ends but is a continuous struggle for advantage. However, Clausewitz also implores strategists understand the answer to the question: what is the ultimate utility of strength?

The obvious answer is that superior strength is not the end but only the means. The end is either to bring the enemy to his knees or at least deprive him of some of his territory—the point in that case being *not to improve the current military position* but to improve one's general prospects in the war and in peace the negotiations. Even if one tries to destroy the enemy completely, one must accept the fact that every step gained may weaken one's superiority—though it does not necessarily follow that it must fall to zero before the enemy capitulates. Hey may do so at an earlier point, and if this can be accomplished with one's last ounce of superiority, it would be a mistake not to have used it....

If one were to go beyond that point, it would not merely be a *useless* effort which could not add to success. It would in fact be a *damaging* one, which would lead to a reaction; and experience goes to show that such reactions usually have completely disproportionate effects.[24]

Lastly, Clausewitz offers that in warfare timidity is most common as well as most deadly:

In reviewing the whole array of factors a general must weigh before making his decision, we must remember that he can gauge the direction and value of the most important ones only by considering numerous other possibilities—some immediate, some remote. He must *guess*, so to speak: guess whether the first shock of battle will steel the enemy's resolve and stiffen his resistance, or whether, like a Bologna flask, it will shatter as soon as its surface is scratched.... When we realize that he must hit upon all this and much more by means of his discrete judgment, as a marksman hits a target, we must admit that such an accomplishment of the human mind is no small achievement. Thousands of wrong turns running in all directions tempt his perception; and if the range, confusion and complexity of the issues are not enough to overwhelm him, the dangers and responsibilities may.

This is why the great majority of generals will prefer to stop well short of their objective rather than risk approaching it too closely, and why those with high courage and an enterprising spirit will often overshoot it and so fail to attain their purpose. Only the man who can achieve great results with limited means has really hit the mark.[25]

Whereas Handel's curve can help us determine where the campaign's culmination point is, it is the S-curve that can help reveal at which time we are at during the campaign. Foster's work on S-curve analysis suitably modified can be used to advantage.[26] Foster provides a series of questions that can assist in the technological warfare campaign. The first thing that must be determined is whether the campaign is set up to detect technological discontinuities. This determination is essential to begin the campaign, and perhaps anticipate its end. In order to properly evaluate and position the campaign, the following questions must be answered:

1. Have you identified the key performance factors that translate most effectively into military utility?

Technology by itself is sterile. To be militarily useful the new technology must provide new or better solution to a military problem. Military performance usually takes the form of an improved capability such as speed, payload capacity, destructiveness, ease of use, and the like. What is necessary here is the effect produced, not necessarily how it is produced. In order to be successful, the technological war campaign must know which performance factor will provide the highest utility and has the most potential to surprise the enemy.

2. Do you understand the relationship between the key performance factors and the key design variables?

Do we understand what the engineers can change in the design of the equipment that will produce improved performance? What are the cheapest and quickest changes available that will produce the desired improved performance? This step is crucial because it translates the military need into engineering and research development avenues. Understanding what may be changed and how easily these changes may be accomplished is essential to planning the campaign.

3. What are the limits of the key design variables?

How much untapped potential improvement still exists in the current technological methods? For instance, the development of the P-51 had begun to push the extreme limits available for propeller-driven aircraft (a key design variable) to produce speed (a key performance factor). The limits of propeller technology were being reached, leading aircraft designers to look to a new method (jet engines) to further increase performance. If there is still a great deal of engineering improvement theoretically left, you may still be comfortably in the ascending area of your S-curve. However, if simple engineering improvements may no longer be an option to provide enhanced performance, it may be time to jump to a new S-curve by researching new classes of technology.

4. Have you identified your competitor?

A technological campaign is worthless unless you have a clear target of your campaign. While technological performance can be improved against all potential competitors simultaneously with a successful innovation, a competitor must be specifically targeted to know how your relative power is changing due to aggressor and conservator actions. The campaign must have a target.

5. Do you know the limits of your competitor's approach?

Does the target of your technological warfare campaign have a different technology than you do to achieve military performance? If so, do your engineers and researchers understand the limits of the technology that they're using? What are the relative merits of your approach versus theirs? The enemy gets a vote even in technological warfare. It's important not only to understand your S-curve but also your opponent's in order to estimate relative power differences.

6. Do you know whether or not your research and development productivity is increasing or decreasing?

Measuring R&D productivity is critical, for it is one of the best ways to estimate where you are along the S-curve. The rate of performance improvement divided by cost (in both time and money) and can give you the estimated slope of your S-curve. A high and increasing slope often indicates a great deal of performance improvement still ahead. However, R&D gets very expensive as a technological limit is approached and a discontinuity is close. A decline (no matter how high the slope) may indicate you are approaching T2 or T3 and should be weary of reaching the culmination point of the campaign and explore jumping to a new S-curve. Foster recommends examining R&D productivity rates over long periods of time to ensure that any apparent effect is real.[27]

7. Can you answer all of these questions for your opponent as well?

Since this is a campaign against an active opponent, and the campaign curves involve both the aggressor and the conservator S-curves and relative power lines, it is imperative to know as much as possible about the adversary's technological structure as well. Intelligence must play a large and vastly important role in the technological warfare campaign.

With these questions answered, we can now look to the specific problems of the technological war campaign and focus on a specific innovation. Doing this requires four types of analyses in order to build a representative S-curve:

1. Identify alternatives to your present approach.

The aggressor must list all identified options available to produce the desired performance enhancements in your key performance factors. All options should be listed in order to choose the best one as well as to anticipate what the conservator might do to counteract the aggressor's advance. This analysis is meant to ensure the aggressor deeply understands the issues surrounding the performance factor that will generate relative power.

2. Identify performance parameters.

Here the aggressor must aggressively study the technologies being used to generate the key performance factors and their state of the art in order to select the easiest ones to improve to increase performance. Performance parameters change over time and looking at the historical performance of the type of equipment in question can help the aggressor estimate what the future performance parameters should look like. Both absolute values and their rate of change can evolve over time and the aggressor should attempt to form a best estimate for both.

3. Identify the limits of technology.

After developing an estimate of the required future performance parameters, the potential limits of each technology should be explored. Does a popular structural material for an engine have a melting point that may soon be reached? Are there fundamental laws of physics that will prevent further improvements along the new technology's S-curve? Once these theoretical limits have been identified, the aggressor may then be able to offer an estimate of the numerical performance limits based on these numbers. With these, an estimated top of the S-curve can be generated and serve as a warning for determining when R&D may begin to yield diminishing returns and T2 or T3 may be reached.

4. Draw the S-curve.

With estimates of the aggressor's and conservator's current relative power and R&D situation and a grasp of the theoretical performance limits of the new technology, an S-curve can be sketched. When done, identifying the objective of the campaign and estimating the relative power change necessary for the conservator to yield to the aggressor's objective must be done to see if there is a good chance the technological warfare campaign may succeed given the potential of the innovation. If the objective is difficult and the technological limits of the new technology not sufficiently advanced, the campaign may fail or have only a very slim window of success. Alternatively, an easy goal mated to a technology offering a quantum leap of performance may not be ambitious enough.

Although these analyses are important, really the most critical pieces of information to determine are the rates of performance increase over cost for both the aggressor and conservator and to determine their absolute performances as well. At the beginning of a campaign, the aggressor's performance increase to cost ratio will be much higher than the conservator's (called the attacker's advantage by Foster). At T2, the conservator's ratio will become equal to and then greater than the aggressor's and the aggressor will enter the zone where he is in danger of reaching the campaign's culmination point. When the aggressor's and the conservator's relative power from the technology has reached parity again, T3 has arrived and the campaign is over. To offer a final quote from Possony et al.:

> In technological warfare, generalship is the key to success, as it always has been in every conflict. The difference today is that generalship on the battlefield is perhaps less important than generalship exercised many years before a battle is joined. This is especially true of the generalship that goes into the design and development of weapon systems. The general who wins the battle is usually the man who held decisive control ten years before the fighting started and who, at the moment of battle, is either dead or retired.
>
> Technological generalship must anticipate strategy, tactics, and technological trends. It must develop weapons, equipment and crews. Such developments must be anticipated in advance of trends.[28]

It is hoped that this model of a technological war campaign will help technological generals and their staffs better anticipate technological trends and develop effective strategies with which to conduct technological warfare. As Holley says, "Sometimes the advantage of a superior weapon is decisive before countermeasures can be evolved. It follows then that the methods used to select and develop new weapons and the doctrines concerning their use will have an important bearing upon the success or failure of armies—and of nations...[29] To exist in a warring world the nation must pick winning weapons; in military analysts will distill every possible lesson ... such weapons will be easier to find and the odds of national survival will go up."[30] Developers of space power must become masters of technological warfare and the technological warfare campaign. But which organization is best capable of developing technology itself? This question will be the focus of the next section.

The Merchant and the Warrior

Financier James Rickards says that to have a complete understanding of modern economics requires us to understand the rise of state capitalism. State capitalism is a new form of the classic mercantilist economic philosophy which eschews free markets and focuses on national accumulation of land, commodities, and gold which the mercantilist believes is wealth. The agents of the mercantilist economies are national governments and state-backed private corporations. In the past, the British East India Company carried the mercantilist flag. Today's mercantilists are the agents of state capitalism, such as "sovereign wealth funds, national oil companies, and other state-owned enterprises."[31]

Rickards accuses the mercantilists of believing that trade is war and insists that mercantilism consists of accumulating wealth at the expense of others.[32] This is a biased view of mercantilism that assumes that a certain amount of a nation's wealth achieved through protectionist measures *deserves* to be accumulated by foreign nations if the foreign nation would get that wealth under free trade. Economist Ian Fletcher believes, rather, that the essence of mercantilism is a soft form of nonideological economic nationalism that believes "a nation's economy should basically be run for the benefit of its people."[33] Regardless, the world of currency and resource wars is the world of the mercantilist, and in order to defend itself in this world a nation must reacquaint itself with the historical tools of the mercantilist. The first, and most powerful, mercantilist tool was a nation's merchants. Economic power rested in the firms a country could produce. These agents are easy to understand and the merchant space service (like The Spaceship Company, SpaceX, and traditional firms like Boeing and Lockheed-Martin) needs no explanation. What we will describe is the critical enabler of the historically understood—but now almost entirely forgotten—nonwarfare uses of a nation's military to merchant success.

Rear Admiral Robert Wilson Shufeldt, United States Navy, was an early naval thinker who opened up United States trade to Korea and whose theories predated Admiral Alfred Thayer Mahan's by only a few decades. His 1878 letter to the Honorable Leopold Morse of the U.S. House of Representatives Naval Committee entitled *The Relation of the Navy to the Commerce of the United States* offers a very clear description of the reciprocal relations between a nation's navy and its merchant marine, a description not surpassed even by Mahan. Admiral Shufeldt offers four ways in which the merchant service and the Navy both strengthen the other.

Firstly, Admiral Shufeldt argues that merchant ships can quickly become military ships if required, provided the merchant ships have certain speed and fighting qualities the military can use.[34] Secondly, and perhaps more importantly than the first, merchant sailors can quickly become military sailors. Shufeldt, lamenting the decline of the merchant marine service in the late 19th century, writes, "During the late war [the U.S. Civil War] the Navy drew from the merchant marine *four thousand five hundred officers*—deck officers and engineers.... During the war *sixty thousand men*, the rank and file of the Navy, came in from the merchant service. Should another war come upon us ... where are these ships, these officers and men to come from, unless the mercantile marine of this country is restored to its former prestige?"[35] Here, Shufeldt proposes that the merchant marine offers a reserve to the Navy in both ships and personnel that can be called upon by the military in times of national security crisis. This is a direct sea power analogy to how economic space power can be transferred into military space power when required by events. Merchant rockets (and other space equipment) can quickly be used to serve military payloads and missions. This is an essential relationship for military space professionals to understand and internalize, but the Navy's effect on the success of the merchant marine is far more enlightening to understanding how merchant space forces can be developed to success.

Shufeldt continues, "But if the mercantile marine is so essential to the Navy it is safe to say the Navy is no less indispensable to commerce. *The Navy is, indeed, the pioneer of commerce.*"[36] He explains:

> In the pursuit of new channels the trader seeks not only the unfrequented paths upon the oceans, but the unfrequented ports of the world. He needs the constant protection of the flag and gun. He deals with barbarous tribes—with men who appreciate only the argument of physical force.... The man-of-war precedes the merchantman and impresses rude people with the sense of the power of the flag which covers the one and the other....
>
> Travel where you may over the boundless sea, you will find the American flag has been there before you, and the American Navy has left its imprint on every shore— no less in peace than in war.[37]

Though a future space military will likely not spend its time dealing with barbarous tribes and other rude people, the harsh environment of space will provide ample scourges from which the merchant space service will need protection. Rapid astronaut rescue and recovery, delivery of emergency supplies, and protection from solar storms as well as predations from foreign interests will be services a nation's military space power agents can provide to its economic space power brethren. But perhaps more directly, the military can blaze trails with which merchants can follow with confidence of success. Shufeldt relates the U.S. Navy's efforts:

> [Matthew Fontaine] Maury, the geographer of the seas, with his wind and current charts, making the paths of commerce plain to the commonest understanding; [Thornton A.] Jenkins, the founder of the light-house system, dotting the coast of America with its lights, buoys, and beacons, now as safe to the mariner as the gas-lighted street to the wayfarer; the Coast Survey, with its unequalled charts and sailing directions for thousands of miles of shore, and bay and river; [Robert H.] Wyman, in the Hydrographic office, watching every discovery of shoal or rock upon the ocean, and warning the somewhat heedless mariner of his danger. Nor should we forget the National Observatory, with its insignificant means and its dilapidated buildings, yet holding its place among the scientific institutions of the world.
>
> All this, while acting as the police of every sea, the Navy has done in the aid and for the aggrandizement of American commerce.[38]

Thus can military forces blaze the trail for the merchants that will come after them. But the Navy had one other contribution it could offer to the merchant marine. In Shufeldt's fourth example of the reciprocal relationship between the merchant marine and the Navy, the Navy can be the training ground for the next generation of merchant seamen. "[The Navy] is educating five or six hundred American boys per annum; many of whom at the age of twenty-one will go into the merchant service, thoroughly disciplined and drilled to become its officers and seamen, taught to believe in the flag which floats over them, and proud of the country of their birth."[39] Naval training can be a skills multiplier in the merchant marine. Today's military space forces can also be used as a springboard for people to learn important skills essential to the merchant space service as well.

The U.S. Navy understood this role the military played in expanding commerce and increasing the wealth of the nation under the mercantilist system of the 18th and 19th centuries, and so did the merchant marine and nation understand the importance of the naval service in this respect. "Do not these facts compensate in a great degree for the expanse of maintaining a navy?" Shufeldt argued. "If they do not, then this nation is a mere myth, and national progress an utter absurdity. I, for one, however, still believe in the inherent greatness of our people. I believe that our merchant marine and our Navy are joint apostles, destined to carry all over the world the creed upon which its institutions are founded, and under which its marvelous growth in a century of existence has been assured."[40]

Here is an explanation of how military and economic sea power are connected together, sharpening and supporting each other, and through this relationship we can see how military and economic space power are equally related. A merchant space service that could harness the power of the Sun to beam cheap and plentiful electricity from space would be able to provide the wealth necessary to maintain a powerful space military. A merchant service that could mine the Moon for rare materials and flank potential adversaries in a resource war through economic space power could provide an unparalleled reserve to military space forces in a national emergency. A military service that could provide a steady supply of trained personnel to the merchant space service, and provide protection and support to commercial efforts, would earn the undying devotion and support of the commercial sector.

The link between military and economic power, as well as the merchant and mercantile services, was a well-established understanding during the mercantilist age. However, with the growth of the free market idea today we have needlessly divorced and forgotten these connections. The military is ignored by the commercial classes, and the military has abandoned all thoughts to commerce in its single-minded mania to "support the warfighter" (see Chapter 2). In order to develop the merchant space service, we must re-embrace the mercantilist understanding of the relationship between military and commercial power, and use this understanding to guide both military and economic policy. This will be the focus of the next section.

Developing the Merchant Service

Admiral Shufeldt was clear in how he wanted the U.S. government to act on rebuilding its merchant marine service: he advised the government to provide subsidies to private companies for developing steam-powered ships with required specifications and beginning steamship mail packet service between strategic routes. With regard to new technology, Shufeldt writes:

> Since the introduction of steam upon the ocean, experience has proved to Great Britain and other commercial powers, that capital will not invest in steam navigation without some security from the Government against total loss. The risk in seeking trade, together with the large investment required to inaugurate it, frightens the capitalist; but let the enterprise obtain a footing, then it continues by virtues of its merits and capabilities, or if it fails for want of them at the end of, say, five or ten years, it no longer deserves support.
>
> *In no other* way can our commerce be re-established or our prestige restored on the ocean.[41]

Shufeldt argues that strategically significant trade routes and technology should be encouraged by government policy due to their inherent strategic utility beyond simple profits. Like today, however, Shufeldt confronted an American distrust of government subsidy:

> Owing to the maladministration of one steamship company the idea of a subsidized line of steam packets has become odious to the American people; but it becomes the legislators of the country to boldly confront a prejudice when the public good clearly demands it, only guarding our legislation in such a manner that the faults or crimes of one company shall not operate to the injury of others.[42]

Shufeldt's arguments are what economists refer to as industrial policy: prescriptions for government action to encourage the development of specific industries of particular interest to a nation; in Shufeldt's case the American merchant ocean shipping industry. Shufeldt argued for government action to support the merchant marine primarily for national defense autarky reasons— that a robust and nationally self-sufficient merchant marine was necessary for the Republic's defense. As military policy, this argument is easily understood. However, he also advocated government action for national wealth reasons. This argument is not so clear because modern economics is primarily dedicated to free trade rather than mercantilist notions. We will discuss this old idea using modern trade theory.

Nations trade because each nation has a natural comparative advantage in producing something other nations want. Nations with comparative advantages in aircraft will trade their aircraft for goods that other nations have a comparative advantage in, such as oranges or tractors. Classical Ricardian trade theory (named after its developer, British economist David Ricardo) suggests that under free trade conditions (i.e., no government controls, taxes, or subsidies placed on trade activities) there will be exactly one economic outcome, known as "equilibrium," where every nation trades according to its comparative advantage and thus maximizes world output and wealth.[43] Under perfect freedom to trade, this equilibrium outcome is both inevitable and benefit maximizing on a worldwide scale. Under the Ricardian free trade model, mercantilist notions and prescriptions like Shufeldt's serve no rational economic purpose: any government action that interferes with an industry to alter its outcome other than the purely free trade equilibrium outcome results in inefficiency and hurts world production.

However, in their 2000 book *Global Trade and Conflicting National Interests*, mathematician Ralph Gomory and economist William Baumol explode the foundations of free trade theory. They conclude that nations are no longer constrained by natural and permanent comparative advantages and can now alter their comparative advantages to fit national strategies.[44] Since comparative advantages can be manipulated through national strategy (abetted by the rise of scale economies—in most modern industries both labor and capital requirements are so resource intensive that the more a national industry can produce, the cheaper they can sell each unit of production; therefore, an industry may be able to be so advanced in its infrastructure that it will be able to undercut any potential competitor because it has a natural monopoly due to its size and ability to produce more cheaply and is relatively safe from foreign competition, and hence retainable by the host nation), there is not one equilibrium solution but now an almost infinite number of possible outcomes to world trade, and some solutions will be better for a nation than others. Under the Gomory/Baumol trade theory, classical mercantilist strategy is given a sound theoretical base. As Gomory and Baumol explain, in a world where multiple trade outcomes are possible, the very best trade outcome for one nation will likely be bad for another.[45] It turns out that trade is war, or at least, a nation's economy can be managed to benefit the nation's people more than free trade would allow.

Shufeldt's advice to subsidize technological advances and the development of new trade routes, as well as train merchant marine personnel through the use of the navy, can be interpreted to be ways for the U.S. government to improve its comparative advantage in the merchant marine industry and form a world trade outcome more advantageous to the United States. In fact, such measures are advocated by the Gomory/Baumol model:

> The means for bringing productivity up to snuff [relative productivity increases being the way to increase comparative advantage] vary. Industries that consist of small companies can be helped by an industry association that gathers information on the most productive methods from around the world. In some cases government or the industry itself can support re-education and training of the workers in the industry. Again, an approach suited to the needs and attributes of the individual industries probably is best, and some industries will prove to be beyond help. But the goal is clear. Industries should, where possible be encouraged and possibly helped to approximate maximal productivity. If government can find effective ways to help this happen, that is in the national interest, not only in the parochial interest of the industry involved....[46]

The economists continue that specific incentives such as "stimulating tax concessions" and infrastructure investments to targeted industries can help nations shape their comparative advantage.[47]

Industries can be supported for two reasons: their necessity for national security reasons and their ability to generate relatively greater wealth (through increased numbers and compensation for jobs or abnormal industry profits) for the nation. Because of the relatively high wages and technical skills required for aerospace ventures, it's a virtual certainty that the merchant space service—the commercial space industry—is worthy of protection for its importance in national security as well as its ability to increase the national wealth of the United States. But this leaves us with the question of just how we subsidize this industry.

Here it is important to realize a critical observation. On a national strategic level it is the national industry—not an individual firm—that is important. In other words, it is the U.S. steel industry, not U.S. Steel, that is the legitimate focus of national attention. Therefore, government subsidy should not fall into the oft-mentioned indictment against industry subsidy of "picking winners."

But strategic interest in the health of the national industry rather than particular firms has a positive component in addition to its negative component advocating against subsidizing a particular firm over others. It opens the possibility of protecting national firms from foreign competition while simultaneously encouraging a healthy and robust domestic competition, further increasing the national advantage. Of even greater interest, domestic competition may, in fact, be more advantageous to productivity than foreign competition. Business strategist Michael Porter writes:

Domestic rivals fight not only for market share but for people, technological breakthroughs, and, more generally, "bragging rights." Foreign rivals, in contrast, tend to be viewed more analytically. Their role in signaling or prodding domestic firms is less effective, because their success is more distant and is often attributed to "unfair" advantages. With domestic rivals, there are no excuses.

Domestic rivalry not only creates pressures to innovate but to innovate in ways that *upgrade* the competitive advantages of a nation's firms. The presence of domestic rivals nullifies the types of advantage that come simply from being in the nation, such as factor costs, access to or preference in the home market, a local supplier base, and costs of importing that must be borne by foreign firms.... This forces a nation's firms to seek *higher-order* and ultimately more sustainable sources of competitive advantage.[48]

As Fletcher says, protecting a nation's industry from foreign predation while encouraging a strong domestic rivalry is probably an even better innovative environment than even unrestricted free trade, as Japan's internationally protected but internally vicious electronics industry attests.[49]

Fletcher argues that a natural strategic tariff—a flat percentage tax on all foreign goods imported into the United States—would go far in protecting America's internal industries. It would be simple, would not pick specific industry firms as winners, and would be remarkably strategic:

Although this is a complex issue, the fundamental dynamic is clear from the fairly obvious fact that a flat tariff would trigger the relocation back to the U.S. of some industries but not others. For example, a flat 30 percent tariff (to pluck a not unreasonable number out of thin air) would not cause the relocation of the apparel industry back to the U.S. from abroad. The difference between domestic and foreign labor costs is simply too large for a 30 percent premium to tip the balance in America's favor in an industry based on semi-skilled labor. But a 30 percent tariff quite likely *would* cause the relocation of high-tech manufacturing like semiconductors. This is key, as these industries are precisely the ones we *should* want to relocate. They have the scale economies that cause retainability, high returns, high wages, and all the other effects of good industries. *Therefore a flat tariff would, in fact, be strategic.*[50]

This strategic tariff would undoubtedly influence and protect the American space industry. In the space industry, certain sub-industries like traditional orbital space lift may come back to the United States as domestic satellite providers use U.S. launch vehicles rather than Russian, Chinese, or Indian launch services. Other newer industries such as space tourism and reusable launch technology may never need locate outside of the country from the beginning under this tariff protection. Let us assume that a natural strategic tariff or a tariff devoted specifically to space technology adequate to protect the American merchant space service from foreign predation is enacted. How, then, do we subsidize the merchant space service itself on a domestic level?

The answer again is to focus on overall industrial health and not pick individual firms for subsidy as political winners. Two ways to improve the domestic merchant space service will be offered for consideration. First, the creation of tax legislation that will declare a tax holiday on profits derived from merchant space activity for a fixed number of years would spur private investment and not unduly favor one domestic space firm over another. The so-called Zero-G, Zero Tax bill passed by the House of Representatives in 2000 and narrowly defeated in the Senate would have foregone federal taxes on all profits made in off-planet economic activity (excluding existing communications and remote sensing satellites) for 20 years is a model for such a generic subsidy to the American merchant space service without unduly preferring one firm over another.[51]

A second attractive subsidy has already been successfully used to generate innovation in the merchant space service: the prize. The Ansari X PRIZE, a $10 million purse generated from private funds to be given to the first company able to fly three people to altitudes above 100 kilometers high in a reusable spacecraft twice in two weeks, was won in 2004. The resulting ship, the *WhiteKnightOne/SpaceShipOne* carrier aircraft and suborbital spacecraft has launched the Virgin Galactic and the Spaceship Company businesses, as well as been the prototype for the first commercial suborbital space tourist spacecraft. The X PRIZE, even seven years later, is the genesis for the new Stratolaunch orbital booster system. Some $10 million was the catalyst for an entirely new merchant space industry. And the X PRIZE team did not choose a political winner. The prize was open to all, and was fairly won by the first team to demonstrate the desired capability, not the firm that was best politically connected. NASA's Centennial Challenges program is an ongoing attempt to leverage prizes to improve national space capability. Similar space prizes for technology demonstrations with significant military or economic potential (such as the capacity to mine lunar titanium or produce an efficient reusable launch vehicle, etc.) could be a very valuable tool in increasing domestic merchant space productivity and should be explored further.

Regardless of the organizations or methods used to develop Grammar Delta technologies, Access Technology will be the single greatest driver of space power development. We now turn to exploring a critical area of space transportation technology that promises to be a quantum leap in space power development—replacing chemical rockets with nuclear power technology.

The Nuclear Hammer and the Path to Impulse

Former Atomic Energy Commission policy expert James Dewar believes that the nuclear rocket is the future of American space power because it is "the bigger hammer." He describes his vision of a new nuclear space program in his book *The Nuclear Rocket*. The nuclear hammer is bigger than the chemical rocket because the nuclear rocket is not limited by heat and weight like its chemically driven cousin, but has potential limited only by man's engineering state of knowledge at any particular time.[52] Nuclear rockets have no theoretical maximum limits—they have an open-ended efficiency envelope with no inherent developmental dead ends. Taking advantage of this essential characteristic can revolutionize human spacecraft design and place us on an S-curve that can propel American access to space to the entire solar system—and perhaps beyond.

Dewar maintains that the nuclear hammer and its technological innovation can transform the space program from the expensive and elitist club of today (there are no blue-collar astronauts) into an inexpensive, egalitarian, and equitable one—a consequence more dramatic, he claims, than any other prior epoch-changing shift in human history.[53] *Star Trek* certainly chronicles humanity's transformation from a race suffering from weakness, want, and inequality into a galaxy-spanning civilization of strength, prosperity, and promise—certainly an epoch-changing event!

This epoch shift will be assisted by the nuclear rocket because this technology will extend man's dominion from 20 miles above the surface of Earth to the end of the solar system.[54] In doing so, it will open vast troves of resources, knowledge, and space for human development, virtually eliminating scarcity of raw materials and allowing the economical exploration and exploitation of an area that dramatically eclipses the old frontier of the terrestrial New World. And just as the old New World opened new prospects to millions of adventurous colonists, so this new New World of the solar system will provide purpose and opportunity to millions of new pioneers. The barriers of energy and distance that kept humanity chained to their home world will be destroyed forever by the nuclear hammer.

The nuclear hammer's strength will be determined, as described above, by its operating temperature and the molecular weight of its propellant. Nuclear thermal rockets can be designed to use the very light hydrogen gas, which effectively minimizes propellant molecular weight and provides the maximum benefit to I_{sp} from the propellant characteristic. This leaves operating temperature as our key to gaining high I_{sp}, and high engine efficiency. Since the nuclear rocket uses the power of the atom, temperature yield can theoretically approach the millions of degrees—literally as hot as the stars—making temperatures available for propulsion practically unlimited and causing I_{sp}s to reach astronomical heights. The nuclear hammer can begin as a handyman's hammer, light but powerful, and mature into an industrial sized jackhammer, making humanity's dominion over the solar system easier and easier, ad infinitum. How it will progress is a powerful story in and of itself, and how it will end is with a system very familiar to even the casual *Star Trek* fan.

A centerpiece of Dewar's book is when he describes the evolutionary potential of the nuclear thermal rocket as a "Nuclear Continuum."[55] His continuum catalogs the developmental path from the rudimentary nuclear rockets available to humanity today to the powerful spaceship drives of centuries hence. This seems fanciful, but consider the maritime steam engine. The 17th-century steamships could barely fight the current of a small river, but today's massive oil tankers use engines that are direct descendants of the boilers on the rickety and dangerous steam vessels of two hundred years before. The nuclear continuum can be envisioned as an S-curve where the beginning allows for low Earth orbit travel comparable to the performance of chemical rockets and ends with performances allowing fast and economical human travel throughout the solar system.

The nuclear engine and the nuclear continuum S-curve have two major limiting factors in their potential, and they are both factors that can be enhanced in an evolutionary way (i.e., they require no major theoretical break-throughs such as discovering hyperspace or completely new physical principles). They are the ability to control and harness high temperatures, and the ability to control the nuclear process itself.[56] The nuclear continuum is a visual description of the engines that may be available as we progress in each of these categories (Figure 3.4).

The nuclear continuum is divided into five classes of engines, all characterized by the nuclear reactor they use (Dewar ignores the nuclear pulse propulsion Orion system as discussed in Chapter 2, for obvious political reasons). The first three (solid, liquid, and gas core) use nuclear fission as their heat source, and fusion and antimatter reactors are concepts that can theoretically provide ever-higher operating temperatures which lead to orders of magnitude increases in I_{sp} to fantastical levels. However, fusion and antimatter reactions are currently beyond our engineering capability. It is important to keep in mind, however, that from an engine perspective we are only interested in higher operating temperatures. We know that fusion reactions yield more energy and can achieve higher temperatures than fission engines, but fusion reactors are only useful if we can use their higher operating temperatures. Therefore, the nuclear continuum assumes that our ability to contain these reactions has advanced far enough to use these nuclear reactions as power plants. Thus, the nuclear continuum models both our mastery of nuclear processes (the "type") and our ability to harness the temperatures produced for propulsion (the "specific impulse" column). Let's briefly explore some of these engines.

The Nuclear Continuum

Type	Specific Impluse(s)	Comment
1.Solid Core	825-1200+	Demonstrated techology
2.Liquid Core	3000+(?)	Reasearch inconclusive as to its practibility
3.Gas core	8000+	Reasearch inconclusive as to its practibility
4.Fusion	Up to millions	Many concepts, but fusion has yet to be deonstrated
5.Antimatter	Up to millions	Extreme radiation may make all concepts impractical or impossible

Figure 3.4 The Nuclear Continuum (Dewar, p. 37).

The Starting Block: Solid, Liquid, and Gas Core Fission

Fission is the process by which heavy (usually uranium) atoms are bombarded by neutrons, causing them to split and release great amounts of energy. Our understanding of fission reactions is sufficiently advanced to make economical and productive use of it practical. Our nuclear power plants and nuclear powered sea ships are powered by fission reactions. Indeed, NASA has even tested a live fission rocket. Fission reactions can produce great amounts of heat compared to chemical reactions and represent a great leap over chemical propulsion. They also have a great deal of potential to expand our space propulsion capability as three different types of fission rocket in the nuclear continuum offer an order of magnitude greater I_{sp} than available with chemical rockets. The fission rocket's three different classes showcase the nuclear continuum's dual representation of the command of the nuclear process as well as the ability to harness its power. The fission reaction will be essentially the same in each of the three classes of engine, but they way we harness it will be fundamentally different as we are able to use more and more of the total available energy from the fission reaction.

For instance, the solid core engine uses a solid core to produce a fission reaction. From 1955 to 1973, Project Rover/NERVA (Nuclear Engine for Rocket Vehicle Applications) produced prototype solid core fission engines that operated at 2000 degrees Celsius and yielded I_{sp}s of 825 seconds—almost twice that of our best chemical engines.[57] At this temperature, the fission core was made of solid materials, hence the name. The 2000° C limit was not due to the maximum available temperature available from fission, but from our ability to harness high temperatures. With four decades of advances in materials engineering, Dewar believes we are now capable of building solid core engines using fission reactions of 3000° C, yielding I_{sp}s greater than 1000 seconds.[58] However, solid core engines begin to hit a snag as we get hotter. Even though fission reactions can burn much hotter, all of our known materials (including the uranium fuel) begin to lose integrity and melt around 4000° C.[59] To use familiar *Star Trek* terminology, we have a containment breach. Fear not, engineers won't ever let as minor a problem as melting engines stop them. Enter the liquid and gas core designs.

If you get water cold enough, it will form ice. Warm an ice cube up, and it will turn into liquid water. Boil it and you will get water vapor, a gas. Nuclear fuel capable of producing fission, such as uranium, works the same way. Therefore, liquid- and gas-core engines operate at temperatures where uranium melts into liquid and turns into a gas. In liquid core engines, the uranium fuel is allowed to melt and is contained in the engine in a pot-like structure that resembles how molten iron is contained in a crucible in steel foundries.[60] Allowing the fuel to melt would allow higher operating temperatures, potentially yielding I_{sp}s of up to 3000 seconds or more. However, fission can get even hotter, and designing an engine that allows the nuclear fuel to vaporize will be able to harness even larger temperatures. Operating at crushing pressures and hellish temperatures, uranium can turn into an energetic plasma gas contained by structures similar to a fluorescent light bulb that can gain efficiencies of 8000 seconds or more.[61] Large scale human colonization of the inner planets becomes feasible at 2000 sec I_{sp}s, and with 8000 I_{sp}s spaceships can economically travel to Mars in a month! Fission rockets can provide all of these benefits.[62] However, once uranium can be used as a gas, we begin to reach the maximum temperatures available from fission. We must then turn to more exotic methods.

The Impulse Engine: Fusion and Antimatter

Fortunately, the power of the atom can provide operating temperatures far greater than simple fission reactions. Fusion reactions, the combination of light atoms such as hydrogen to produce heavier elements like helium, can result in temperatures of literally millions of degrees. Indeed, fusion reactors power the universe in the form of stars like our Sun—all massive fusion reactors. Controllable fusion is just beginning to be matured by humans, but we already have the ability to start uncontained fusion reactions in the lamentable hydrogen bomb. Even more advanced than fusion reactions is the theoretically perfect energy conversion of matter-antimatter annihilation, where matter comes in contact with antimatter, resulting in a complete conversion from mass into energy. Humans can produce antimatter in very small quantities today and the annihilation reaction is well known, if not yet well understood. Thus, fusion and antimatter rockets are not the realm of fantasy but reasoned (if futuristic) speculation. Either of these engines can theoretically produce temperatures in the millions of degrees in any scale, and hence I_{sp}s in the millions of seconds.

It is in these futuristic but foreseeable fusion engines that propulsion technology intersects with *Star Trek*, a commonly understood vision of mature space power. According to the *Star Trek Encyclopedia*, the impulse drive is a "spacecraft propulsion system using conventional Newtonian reaction to generate thrust. Impulse drive is powered by one or more fusion reactors that employ deuterium fuel to yield helium plasma and a lot of power.... Normally, full impulse speed is one-quarter the speed of light."[63] Does the fusion rocket on the nuclear continuum match this definition? The fusion rocket relies on Newtonian (action-reaction) thrust. Multiple deuterium-helium fusion reactors can be used to power multiple rockets or nozzles. Project Daedalus, a fusion rocket paper design experiment conducted by the British Interplanetary Society between 1973 and 1978, found its ship could reach 12 percent of light speed.[64] It would stand to reason that theoretical fusion rocket advances that could net another ten percent of the speed of light would not be insurmountable.

Can there be any question that the impulse engine of *Star Trek* fame is a fusion rocket, and simply a very advanced (but maybe not even the most advanced) type of nuclear rocket on the nuclear continuum? The fusion rocket matches all of the requirements of not just meeting the impulse drive's specifications, but actually being the operating mechanism of the impulse drive itself. This example is very convincing that following the nuclear continuum will net humanity the impulse drive, and that the continuum is a very important road map towards a mature space power propulsion future.

We must always keep in mind that the nuclear continuum is not something we should expect to travel completely in a few years or decades. Many steps in the continuum will be beyond our engineering capability for many years and perhaps many lifetimes, just as the first steamship builders couldn't comprehend our nuclear warships of today. But we cannot let that discourage us. There is plenty we can do with engines at the beginning of the continuum, and the sooner we begin the path, the faster we will reach maturity. We have the map and we have the first engine.

Missions and Cutters

Assume that the United States forms the United States Space Guard (modeled after the U.S. Coast Guard) in response to the need of a new military space organization presented in Chapter 2. Near-term missions such as astronaut rescue, emergency repair, and debris mitigation, as well as midterm missions like orbital security and planetary defense (against Klingons or asteroids!) will require ships that can fly to many different places with little or no notice. Indeed, Space Guard ships will need to reach anywhere that humans or (to a somewhat lesser extent) their machines are operating in space. The Space Guard will never know whether an emergency call for assistance will come from a punctured and leaking space tourist hotel in low Earth orbit or ferry supplies to an exploration crew on the Moon whose supply of food has fallen to the bottom of a crater by accident, but Space Guard ships will have to be able to reach either distressed group. Nudging away an asteroid approaching Earth a little too close for comfort may take ships on far trajectories through much of the inner solar system. Massive rescue operations to relieve a colony on Mars, much like the U.S. Coast Guard's 19th-century operations in remote frontier Alaska, will require space ships with great endurance and flexibility.

Unlike today's spacecraft, which are designed to operate only at specific altitudes and can only change positions with great difficulty and massive cost, Space Guard cutters—crewed spaceships designed to respond where needed almost immediately and be prepared for almost any contingency—will by necessity need to be able to race to the Moon or geosynchronous orbit (both very different missions) without significant difficulty or alteration of operation. Instead of being designed for one specific purpose and being stripped down to the bare essentials to accomplish that mission, the cutter will need to be able to perform many different missions and be designed to accomplish its highest endurance mission. Therefore, the cutter will have a great deal of "redundant capability" that can be called upon when needed but is not necessarily used in normal operation. Rather than designing our cutter to operate only in a narrow band of limitations, we must begin to think in terms of maximum efficient operational range like we do with seafaring ships or aircraft. Instead of designing our ship to operate only at a 200 kilometer altitude orbit, we must design it with an operational range such as reaching lunar orbit from low Earth orbit and back on one tank of gas—at minimum—but based on our best (most powerful, most efficient, etc.) engine rather than one specifically designed only to go to the Moon and back without refueling. This switch in thinking will be a major improvement and dramatic change to contemporary space mission design practiced by NASA and the Space Commands.

Systems Engineering and Missionitis

Current engineering practice for space mission design is to use the systems engineering process. The first step in the systems engineering process is to define the mission—the problem you wish to solve. For instance, if you wanted to build a backyard deck to enjoy sunny summer afternoons, the "right" systems engineering approach would be to resist driving immediately to the hardware store, buying a load of lumber, grabbing the power saw, and starting to cut. The systems engineering approach would be to think before you act and start by carefully defining what it is you really need.[65] Just as with building a deck, before we start building a spacecraft we must think long and hard about what we really need and what problem we are trying to solve in order to know exactly what we need to build to accomplish the mission so we don't waste resources, be they lumber, titanium, or money.

Thinking before acting is always a good decision, and as far as systems engineering goes to drive this sentiment home, so much the better. However, this systems engineering first step demands much more than simply warning us to think before we act, and it is the deeper ramification that can lead one astray. Not only does systems engineering demand thinking before acting, it also demands that we focus on "exactly what we need to build to accomplish the mission." To borrow a nautical phrase, "Here, thar be dragons!"

Ruthlessly focusing on the narrow mission of a single system can dramatically cut the cost of developing a system to fulfill that mission, but there are costs associated with this minimalism. The costs come in a system's redundant capability. Particularly in space engineering (which in many ways gave birth to the field of systems engineering), spacecraft due to their complexity and expense are often only capable of doing one narrow thing—accomplishing their designed missions. Ask a spacecraft to do anything else, say, try to use a communications satellite as a satellite navigation beacon or fly the space shuttle to geosynchronous orbit, and it quite simply cannot. It would often be cheaper to design an entirely new system to accomplish the new mission than to retrofit existing equipment. Systems engineering tends to produce equipment that can only perform very narrow missions and which has a hard time being used for new purposes in new ways.

The problem is that the mature missions of many systems and vehicles often differ considerably from their original stated goals. Take the U.S. Air Force F-16 Fighting Falcon as an example. The F-16 was originally designed as a low-cost air superiority fighter to augment the more capable but very expensive F-15 Eagle. However, after three decades of use, the USAF F-16 fleet's main role is as a ground attack and close air support aircraft—a very different mission. This mission change evolved in large part due to the end of the Cold War, which made massive aerial dogfights for air superiority highly unlikely. Thus, circumstances forced the F-16 to accomplish something it was not originally designed to do. Thankfully, the redundant capability inherent in its air superiority role was enough to allow it to function as a type of light bomber.

Air systems have a maturity that allow them a certain ability to adapt to new conditions. A bomber is mostly a transport plane specialized for holding bombs. However, if a B-1 bomber was designed in ruthless systems engineering fashion to take off from Missouri and drop two specific types of bombs in a custom-fit bomb bay on Moscow while flying exactly 3000 feet above the ground at all times and nothing else, it would not be of much use in a war in Afghanistan. This is how exacting mission requirements are for many space systems. No wonder a satellite or spacecraft is not much use for anything other than its designed mission.

We must remember that regardless of how rational the systems engineering process seems, engineering hasn't always used this mission-centric approach. To stay with the flying motif, take fighter aircraft in World War I. In the technological war to build the best fighters, systems engineering was most certainly not used. At the risk of oversimplification, instead of designers determining a mission requirement to have a climb rate of X, a top speed of Y, and a turn radius of Z and design an aircraft engine to deliver that performance, designers instead took the best available aircraft engine, the state-of-the-art ever increasing, and built a plane able to take advantage of the engine's power and were content that they had the best climb rate, top speed, and turn radius possible. The Lincoln Liberty aircraft engine is an example of a very powerful engine which powered many World War I bombers for the Allied powers as well as postwar observation and transport aircraft. This reuse of technology allowed for incremental improvements in both engine and airframe technology. The Liberty engine was not designed for a specific mission, but rather to be the most advanced and efficient engine possible to provide power to many different aircraft intended for many different missions.

In general, systems engineering can be said to build the engine to suit the vehicle. If the mission requires an engine of a certain power, systems engineering demands an engine with that power level be designed and any additional power is a simple redundant waste. Alternatively, you can also build the vehicle around the engine. An alternative to systems engineering can be to take the best engine available, design the vehicle to take maximum advantage of the state-of-the-art system, and find missions that the new vehicle can accomplish. This approach can be derided as "if you build it, they (customers) will come," but it has a great strength in being able to have redundant capability and adapt to the changing state-of-the-art much faster than the frail and narrow mission vehicles of systems engineering.

Therefore, systems engineering is not necessarily the sole or most rational way to design space vehicles. Indeed, James Dewar calls systems engineering's narrow focus on the single task "missionitis" and does not consider its rise a good thing. Dewar attributes "missionitis" not to systems engineering but to the science advisors of President Kennedy (though they may have been little more detrimental to space power than President Eisenhower's, noted in Chapter 2). The president's advisors advocated a "mission first" ideology that approved a mission first and then built the infrastructure and equipment necessary to accomplish that mission. Unfortunately, this mission-first mindset was primarily intended to kill any program with which they disagreed: disapprove a mission, or make a mission too difficult to achieve in the allowed time or funding budget, and the entire program could be terminated.[66] Thus, proposed missions would need to "promise the stars" to achieve public support (or the support of presidential advisors), but when they failed to meet overly stringent resource constraints and could only "reach the Moon," they were cancelled as failures. This political trick has now morphed into the malignant "missionitis": the prevalent belief that a mission must exist before any technology development can or should be implemented.[67] This belief, though seemingly innocent and rational, yields some very dangerous and unfortunate consequences. Dewar is very insistent that mission-based development is a dangerous way of doing business. "[I] categorically state that it is highly dangerous and irresponsible to assign missions to a new technology without knowing its handling capabilities and without having a management team fully competent in its operations."[68] Specifically, in terms of space engine design, "missionitis" can be described as follows:

> Technically, "mission-itis" thinking focuses on a particular engine design for a specific mission, with two bad results. First, it produces tunnel vision and hinders

seeing a fuller sweep of engine development, as it centers on one design [of a certain efficiency], making it difficult to see that [more efficient engines based on the original's design] can follow quickly. This may not seem important, but [t]his turns space flight economics topsy-turvey [*sic*]. Second, it fosters a negative mind-set, to dwell on the problems and believe they will be long, difficult, and costly to solve.... It [is] dead-wrong—round-peg in square-hole thinking.[69]

These consequences are profound. Imagine, as an example, a *Star Trek* scenario in which Zephram Cochrane's mission was not to break the warp barrier, but to get to Alpha Centauri from Earth in one week. Confronted with this problem, he would rightly deduce that Warp 1 (light speed) would still take his ship 4.4 years to reach even our closest neighbor star and would not accomplish the mission. Instead, he figured that he would need at least a speed of Warp 4 (4^4, or 256 times the speed of light and Warp 1) to get to Alpha Centauri in a week. Anyone can see that as profound a propulsion breakthrough as breaking the warp barrier would be, Cochrane's mission of a week-long interstellar travel would be all but impossible! Even decreasing mission requirements to the star in a year would still make a necessary mission speed of Warp 2+. In this scenario, the two "missionitis" consequences are very revealing. If Cochrane succumbed to "tunnel vision" in his development of the Alpha Centauri engine, he might completely ignore the fact that the most profound breakthrough would be crossing Warp 1, and that refining the engine to produce greater speeds would become much easier after the initial barrier had been broken. Also, he (or his funding sources) would probably become disillusioned and abandon the project entirely because they could not conceive how they could go from sub-light speeds to 256 times the speed of light in one great leap of technology development. They would be blinded to the incremental development potential of the critical warp drive technology and would more than likely throw their hands up in desperation—leaving the warp drive doomed to end its development as a technical paper or scribble on a bar napkin. For a more conventional historical example, Chuck Yeager's Bell X-1 flight is one of the most celebrated feats in aeronautical history for breaking the sound barrier, though Mach 1 didn't give us significantly more mission capability than any number of fighters that could already approach Mach 1. Its critical contribution to engineering was a propulsion breakthrough, not a marginally faster new fighter, bomber, or transport.

An alternative to "missionitis" that recognizes the utility of breaking barriers without being tied to a mission can be termed capability-based development. Capability-based development identifies a technology that promises significant advantages over current methods and focuses on developing that new technology with the goal of securing new skills and capacity to accomplish things that weren't possible with current technology. Development consists not only of developing a workable engine, but to build an entire support infrastructure and industry aimed and supporting and advancing the technology itself. Much as trains need railroad tracks, automobiles need roads, and telephones need poles (or cell towers), all new technology needs supporting infrastructure to be viable.[70] Capability-based development is a systematic approach where progress is incremental and market-driven—not simply obsessed with a "single shot" mission. It is inherently adaptable by design to changing conditions, such as introducing new cost-saving technologies or surprise technological advances, and revels in developing better and faster later generation systems, phenomena "missionitis" thinking often does not or cannot account for.

Advantages of capability-based development over mission-based development are numerous but often ignored in engineering circles today. Former Atomic Energy Commission commissioner James T. Ramey argued:

> [P]romising technology development efforts should proceed through the prototype stage where something is built for full field testing and evaluation. That allows the nation's leaders, or corporate officers, to make more informed decisions on proceeding [with development of the technology in question], yet it avoids strangling promising technologies with mission requirements, which may become apparent only after full-field testing. At the same time, it keeps the budget under control, and it begins infrastructure development to manage and operate the new technology.[71]

Further advantages of the capability-driven approach are the ability of gaining much-needed experience with smaller engines first, as the U.S. Navy's experience with the relatively small and simple engine of the first nuclear submarine, *Nautilus*, allowed the infrastructure to develop that would one day build and operate very large and advanced reactors in modern nuclear submarines such as the *Los Angeles* and *Seawolf* classes.[72] Very important to remember is that even though capability-based development has no initial *missions*, it most certainly has *objectives*, which allows for identifiable performance objectives and is still conducive to good project management.[73]

These two approaches of technology development are sometimes called "mission pull" for missionitis thinking, and "technology push" for capability-based thinking. Though mission pull will always be a large part of development, the capability-based approach will be critical to developing mature space power and thus deserves a respected position in our engineering and program management tool kit. Why is capability development so important to a mature future?

The first reason is that we do not yet have a firm grasp of all possible missions that would be possible or profitable in space. Because most of our thinking on the human future in space is more speculation in fact, focusing on missions only leads us to possibly be too myopic where vision is critical. Missionitis also tends to be a closed system: the mission either works or it doesn't. Capability-based approaches are inherently open systems: the expansion of ability is limited only by the maximum possible capability of the technology in question, not the human-imposed constraints of a particular mission. Capability-based approaches also tend to encourage multipurpose ships and equipment. Since we do not know of everything we should do, or can do, we focus our ability to be able to achieve many new capabilities that allow us to perform many missions whenever any profitable missions are positively identified. For instance, just as British Royal Navy sloops meant to be men-of war turned out to be excellent exploration ships for the Pacific Ocean, a crewed lunar transport ship could very quickly and easily be retrofitted for a manned expedition to Mars, preferable to the alternative of designing a new Martian exploration craft from scratch.

The capability-based approach is also very important because space enthusiasts must understand that even though powerful starships such as the various *Enterprises* are still many decades or centuries in the future, that does not mean that we cannot do anything to put ourselves on the path to their development now! In fact, incremental advances along the nuclear continuum S-curve will greatly advance space power ability and access until humans will be able to travel throughout the solar system regularly and economically. By embracing the nuclear continuum, American space power can connect to its proud maritime and nuclear histories by embracing a historical name for the first NTR space cruiser: *Savannah.*

The Three Savannah*s*

It seems appropriate, though somewhat clichéd, to call the first nuclear rocket powered spaceship *Prometheus.* Naming the ship after the Titan who stole fire from the gods and gave it to humans in Greek mythology does have some symmetry to nuclear fire opening up the solar system to the human race. It is so compelling that the short-lived early 21st-century NASA program to reopen nuclear propulsion research was called the Prometheus project. However, maritime history provides a much more appropriate naming scheme for the ship as history informs us of a similar epoch-breaking event in the history of sea transportation: the travels of the merchant ships named *Savannah.*

Savannah is arguably the most hallowed name in maritime technology advancement. It is known for two ships: the SS *Savannah* of the 19th century that was the first steam-powered ship to cross the Atlantic Ocean, and the NS *Savannah* (namesake of the original), the first nuclear-powered merchant ship. Making the name especially worthy for use on a nuclear spaceship is that these ships were designed for peaceful commerce, not warfare. There was no ocean-spanning steamship before SS *Savannah*, and even though NS *Savannah* was not the first nuclear powered sea vessel (USS *Nautilus*, the first nuclear submarine, is perhaps the most famous) all of its predecessors were naval ships whose primary mission was fighting and destroying the enemy.

The most fitting reason to continue the name *Savannah* into the new sea of space is what each ship has and will accomplish. One *Savannah* freed sailors from the tyranny of the wind and tides and allowed our ships to achieve mastery of the sea using steam mechanical power. The later *Savannah* introduced nuclear-powered sea travel for peaceful purposes. The next *Savannah* will follow in their footsteps to inaugurate the age in which humans can conquer our reliance on the tides and currents of orbital mechanics (gravity gradients and Hohmann transfers) and truly open up all the solar system with speed, agility, and economy. Before we explore the future, let us first gaze back into the past.

SS *SAVANNAH*

The steamship *Savannah* is a testament to the ingenuity and enterprise of free private citizens to explore, conquer, and innovate with peaceful intentions. In the early 19th century, small steamships were being experimented with for use as river vessels. However, in 1818 Captain Moses Rogers, a pioneer of steam navigation, convinced the Savannah, Georgia–based shipping firm Scarborough & Isaacs to retrofit a small packet being built in New York as a transatlantic sailing ship with a steam engine to inaugurate the world's first transatlantic steamship service. Thus, the SS *Savannah* was born.

Savannah's first transatlantic voyage was a "laudable and meritorious experiment," although many observers of the time ridiculed the small ship with a black-smoke-spewing 90-horsepower steam engine and large side-mounted 16-foot long paddlewheels as nothing more than "Fickett's [the shipbuilder's] steam coffin."[74] The world's first-even transatlantic, and transoceanic, steam-powered voyage lasted from May 24 to June 30, 1819, as *Savannah* traveled from her home berth in Savannah Harbor to Liverpool, England. In a voyage lasting a bit over a month, *Savannah*'s steam engine was used about 80 hours.

Though *Savannah* used far more sail than steam power during her only voyage (she was lost off of Long Island, New York, in 1821) across the Atlantic, the die had been cast. Sailing vessels would continue to ply the Atlantic Ocean for many years (indeed, it would take almost 60 more years for the sails to be removed from the steamship), but the Age of Sail began to disappear in favor of the Age of Steam on the ocean when *Savannah* docked at Liverpool. *Savannah* caused Europe to take notice of America's maritime advances and the United States as a technical power.[75] She also compelled Britain, and later the rest of the sea powers, to embrace the steam engine.

Savannah was truly an epoch-changing event in the history of sea transportation. She began the journey that would eventually free men from the tyranny of the wind and current when traveling across the sea. She took the first step on the path to solidifying mankind's dominion over the sea, placing the oceangoing ship on a radical, and rewarding, new path. Almost 150 years later, a new *Savannah* would complete the journey.

NS *Savannah*

The nuclear ship *Savannah* is a testament to the ability of societies to take horrible engines of war and turn them into promises of a prosperous peace. On April 25, 1955, President Dwight D. Eisenhower announced that the United States would build a nuclear-powered merchant ship as part of his "Atoms for Peace" program, a project intended to beat the atomic sword into a nuclear plowshare.[76] How natural and fitting, then, that this ship which would introduce "nuclear power to the commercial sea routes of the world, should be named the NS *Savannah* after the ship which introduced steam to world shipping!"[77] Thus, another *Savannah* was put to sea.

NS *Savannah* was launched on July 21, 1959, at a cost of $47 million, $28.3 million of which was for the fission reactor and uranium fuel alone.[78] Her first test run was on March 23, 1962, near Yorktown, Virginia, and she reached her home port of Savannah, Georgia (near where the SS *Savannah* berthed), on August 20.[79] In her ten years of merchant service she was only one of four merchant nuclear ships ever built, proving that nuclear power could be used for peaceful purposes. She proved nuclear power could be used safely and usefully on vessels other than those dedicated to war. She could travel over 300,000 nautical miles on one load of uranium fuel, eliminating the need for millions of gallons of diesel fuel to travel across the ocean.

NS *Savannah* built upon SS *Savannah*'s technology to extend man's dominion over the sea. Both were steam-driven ships, their difference only being how they generated the heat for the steam. Almost two centuries spanned the steam continuum from the *Savannah* to the *Savannah*. SS *Savannah* pioneered the steam engine using coal, and NS *Savannah* pioneered the replacement of fossil fuels with nuclear fuels. Now is the time for a third *Savannah*, intended for the new sea of space. She will break man's chains to the Earth, as well as introduce nuclear power to replace chemical propulsion in space. In doing so, she will be the first identifiable Starfleet vessel.

USSGC *SAVANNAH*

The United States Space Guard Cutter *Savannah* will honor both of its seafaring namesakes by advancing nuclear power into space and trailblazing the engine that will allow free societies to expand peaceful activities into space, as well as providing protection and support to private space efforts—essentially supporting the best of both worlds of private and government action to guard and develop the space lanes. USSGC *Savannah* will be the ship that pioneers the use of nuclear power for space travel. In doing so, she will be the grandmother of all Starfleet vessels that come after her.

USSGC *Savannah* will be a Space Guard vessel intended to be an all-purpose emergency response ship performing all Space Guard missions. She will be designed to spend her entire operational life in space; she will not take off and land on Earth for every mission like the space shuttle. She will be designed to be based in a low–Earth, low inclination orbit, and from that orbit will need to be able to reach any orbit around Earth and perhaps even be able to burn to the Moon. She will be commissioned into service well before the middle of the 21st century, and will serve as the flagship of the Space Guard's fleet, the first of a new type of high endurance (long range) cutter that will protect and defend the expanding human presence in space.

To design the *Savannah* in this book would be inappropriate, but we can determine a few of her characteristics to make it a useful Space Guard cutter as well as a prototype Starfleet vessel. *Savannah* will be permanently crewed, with perhaps 5–10 officers and men. She will be capable of responding to a variety of emergency situations, demanding that she be able to accommodate perhaps dozens of people or cargo rescued from space disasters. Due to her requirements of both speed and size, only a nuclear rocket will suffice to power her. Thus, *Savannah* will pioneer the use of large-scale solid core nuclear engines for permanent crewed use. Because she will be manned constantly, operate only in space, and be able to reach many different places and perform many different missions, she will cease to be a scaled down, "missionitis" one-trick pony and instead become a powerful multimission space vessel that will lead the march to a human dominion over the solar system with manned spacecraft following the nuclear continuum to the end of the solar system. *Savannah* will be commanded by the military service U.S. Space Guard officers and crew.

In performing her missions, she will be exactly like the vessels in *Star Trek* in every way except for the advanced technology. Her Space Guard crew may be the beginning of a *Star Trek*–like Starfleet.

Words matter, and the designation of cutter is important. *Star Trek* unabashedly uses naval terminology to describe its ships. The Constitution-class starship *Enterprise* is considered a heavy cruiser, and other ship classes are referred to as light cruisers, dreadnoughts, destroyers, and frigates (in generally accepted non-canon sources if not in canon itself). These designations imply that the fleet comprises warships. In the future of *Star Trek*, the Coast Guard–like lineage of Starfleet is accepted without question. Even though Starfleet is a military organization, the first duty is universally acknowledged as peaceful exploration. In contrast, today and in the near future when the nuclear space ship *Savannah* can be built, the struggle for the identity of the military space service will still be highly contested between those who would see the space service adopt the warlike navy model (assuming that all parties would even agree on the desirability of a maritime model for the service at all) and those who would argue for the peace-driven Coast Guard model. Though some might dismiss the debate as meaningless semantics, it is very possible that the designation of *Savannah* as either a cruiser or a cutter would be a major victory for either the Navy warhawks or the Coast Guard humanitarians, respectively.

Why would the cutter *Savannah* be a victory for the Coast Guard Starfleet proponents of the Space Guard? The answer is because the word "cutter" has a specific history as a military vessel with a primarily peaceful mission. The word "cutter" originates from the British Royal Navy. Their definition of cutter was a small warship of 8–12 cannons in a time where Ships of the Line (recall the *Star Trek* calendar of the same name) often had 70 or more cannons. Even though this definition began as a military term, in regular usage cutter began to mean any vessel (including armed vessels) of the British Royal Customs Service involved in law enforcement duties, not in warfare.

In modern usage, the term "cutter" now means any seaworthy vessel used in law enforcement duties. The U.S. Coast Guard has adopted the term as the designator for its ships over 65 feet in length. Even though some Coast Guard ships approach the combat capability of some of their U.S. Navy sisters, and compare quite favorably in armament to most of the front-line warships of many world navies, they nevertheless retain the designation "cutter" as a constant reminder that, although they are military ships that can and will fight if called upon, their primary mission is always one of peace. In order to secure the U.S. Space Guard's institutional commitment to the militaristic but unwarlike vision of Starfleet, it is vital that this corporate vision is expressed in the name of its ships and especially its first nuclear ship, USSGC *Savannah*.

The USSGC *Savannah* will be only the first attempt at harnessing the power of the nuclear rocket to develop ships that will begin to look like the ancestors of the powerful *Enterprises* of *Star Trek*, much as today's powerful supertankers, battleships, and aircraft carriers can trace their lineage to the tiny and underpowered SS *Savannah*. USSGC *Savannah*'s importance will be its taking of the first step into a larger and fundamentally changed world in which *Star Trek* will be that world's fully mature expression. And this new world will not be a technological change as much as a change in perspective. Dewar again says it best when describing the importance of the nuclear rocket:

> This psychological transformation [allowed by the nuclear rocket] is quite impor-
> tant. We mentally will cease to think of [the solar system] as a vast and dangerous
> abyss because it will increasingly shrink in our minds, from a vast Pacific Ocean as
> with chemical rockets to a *Mare Nostrum* with the solid core [nuclear rocket], our
> sea as the ancient Romans called the Mediterranean ocean, then a great lake, then
> a large pond, and ultimately just a nuisance puddle. In other words, progress along
> the nuclear continuum will cause a mental or psychological shift in which the solar
> system's time and distance dimensions are increasingly less forbidden while, at the
> same time, our sense of personal ownership of and dominion over the bodies of
> the solar system increase.... Our successors in a future century may include the cap-
> tain of a fusion-powered spaceship who views the now demoted planet Pluto as
> only a "planetary" warning to slow his ship down as he returns from a venture
> beyond, and not as the outermost "planet" of our solar system, which is our view
> of it now.[80]

Is Dewar's future not that of *Star Trek*? Does his description of the ship captain traveling past Pluto not perfectly fit that of a Starfleet captain return-ing to Sol on impulse engines? USSGC *Savannah* is the first step to begin the technological development that allows the psychological transformation from today's space program to a Starfleet future. Indeed, there is no significant leap from today to *Star Trek* in intra–solar system propulsion technology that can-not be bridged by the Nuclear Continuum. USSGC *Savannah*, thought of in *Star Trek* terminology, will be the first impulse-powered human space ship!

It's not impossible to imagine that somewhere in the vast Federation Starfleet or Merchant Marine, the starship NCC-18181959 (the launch years of the seafaring *Savannah*s) USS *Savannah* is an advanced propulsion tech-nology demonstrator plying the space ways among hundreds of star systems. Her missions are either transporting vital cargoes between worlds, exploring uncharted areas of the galaxy, or protecting Federation citizens from the dan-gers of space. In the captain's stateroom, as is custom, are images of the ships which bore its name in times past. Among the models or drawings are three vessels: an ancient side-paddled steam ship with a curved smokestack, an elegant but still ancient white seagoing vessel with no smokestack at all, and the first true impulse-driven space cutter ever developed—all bearing the proud name *Savannah.*

This excursion into nuclear rockets and science fiction helps illuminate the fact that with proper technology development, management of the Grammar Delta of Space Power may give birth to a space power future in many ways like our most exotic fantasies of space travel. However, we are not currently on this trend towards mature space power. In order to better understand our current trajectory, and how we must change, we shall now discuss the requirements pull and the capabilities push problem in today's military acquisition environment.

The Joint Force Commander as Customer: An Innovation Case Study

A key Schumpeterian insight into economic development is that development (innovation) *does not* originate from the consumer (demand) side of the economic equation, but rather from the producer (supply) end. Schumpeter says:

> To be sure, we must always start from the satisfaction of wants, since they are the end of all production, and the given economic situation at any time must be understood from this aspect. Yet innovations in the economic system do not as a rule take place in such a way that first new wants arise spontaneously in consumers and then the productive apparatus swings round through their pressure. We do not deny the presence of this nexus. *It is, however, the producer who as a rule initiates economic change, and consumers are educated by him if necessary; they are, as it were, taught to want new things,* or things which differ in some respect or other from those which they have been in the habit of using.[81]

Space power development, a subset of economic development applied to space activity, is also primarily driven by the producer, not consumer. This fact has many ramifications for space power developers—especially in military circles.

Since the Goldwater-Nichols Defense Reorganization Act of 1986, decisions regarding force development have been mostly in the hands of the Joint Force Commanders (JFCs), as they document their "needs" to the various services through service-culture deracinated organizations such as the Joint Requirements Oversight Council. Supporters of Goldwater-Nichols pride themselves on having curtailed rampant "interservice rivalry," which they deem as an unmitigated evil. However, making force development decisions based mostly on the "needs" of the JFCs contains a critical flaw that may prove fatal in the future, a flaw that can be exposed through business theories of innovation.

From a business viewpoint (fully supported by military terminology), Goldwater-Nichols placed the JFC as the "customer" and the "organize, train, and equip"–responsible services as "producers." However, the act went further by giving overwhelming force development power to the JFCs and their joint apparatchiks, effectively eliminating service cultures as forces of innovation. Using economic terminology, Goldwater-Nichols mandated a customer-driven defense force development process. Here lies the problem. It's a well-understood business axiom that producers whose strategy is to focus solely on stated customer needs are incapable of innovation and will ultimately fail to those less beholden to customer opinion. This observation has been documented for over a century.

Schumpeter wrote, "It is, however, the producer who as a rule initiates economic change, and consumers are educated by him if necessary; they are, as it were, taught to want new things." Business and military innovation are not so different that they don't follow the same logic and behavior. Therefore, accepted business theory indicates that the JFC (customer) driven force development scheme will tend to retard profitable force development innovation.

Space, an environment never having developed a mature, service-oriented culture (cultures that may be said to have strived for producer-driven innovation, which Goldwater-Nichols supporters falsely accused as being nothing but interservice rivalry), may be where this innovation flaw has been most apparent. As late as December 2011, the mission of Air Force Space Command (AFSPC) was "to provide an integrated constellation of space and cyberspace capabilities at *the speed of need* [emphasis added]." The need being, of course, the JFC's stated need. A more customer-driven mantra is hard to imagine.

AFSPC's customer-driven mentality is showcased in the Joint Operationally Responsive Space (ORS) construct. Many JFC needs can be served by space assets, including positioning, navigation, and timing (PNT), imagery, communications, and friendly forces (blue force) tracking, among others. Currently, these services are provided using expensive and slowly procured systems with global effect. ORS promises to the JFC no truly new theater effects, but merely aims to offer more tailored services in spans of days-to-months rather than years. Additional space capability, under the ORS strategy, is to be provided by "micro" or "nano" satellites delivered by fast but small space launchers. In this manner, ORS clearly reflects common customer wants as told to producers: customers want the same thing, only cheaper/faster/better/smaller, and so on. Stated customer wants are often merely incremental improvements to existing capabilities rather than true innovations. ORS's promised "innovations" are only incremental improvements to our national space capability, examples of customer driven innovation strategies' classic behavior.

While most may say that the one operative descriptive word for ORS is "fast," an argument can be made that the best word is "small." ORS intends to become responsive by making satellites as small as possible to decrease costs and lift requirements. Unfortunately, the new commercial space industry's rallying cry is "big." Space's most innovative new companies are developing space stations and large rockets to increase the human presence in space, projects diametrically opposed to ORS's small technology response to the JFC's static needs. Interestingly, SpaceX's *Falcon I* launch vehicle is touted as one of the few ORS technologies that has been built and flown (SpaceX being a beneficiary of ORS money). However, founder Elon Musk's motivations for starting SpaceX was to help found a human settlement on Mars—a space enthusiast rationale—rather than to compete for defense ORS contracts. Additionally, the *Falcon I* fit ORS needs, but the rocket's real purpose was to be the technology and operations demonstrator for the much larger *Falcon 9*, a launch vehicle meant to assist "big" space as a cargo and astronaut carrier. Indeed, SpaceX is a vital example of a nondefense, producer driven innovation (the Falcon class launch vehicle) filling a JFC need.

However, imagine SpaceX's "big" space vision completed: a robust orbital infrastructure and economical heavy lift capability effective enough to support a Martian colony. It's unimaginable that such an infrastructure would be unable to achieve ORS goals in addition to its main function. But this ORS enabling solution would not be "small" as the JFC-demand driven ORS strategy imagines—it would be "big" in the fullest sense. By being producer-driven rather than consumer-driven, the JFC could get his ORS needs met with a true innovation, a robust space infrastructure, rather than the comparatively limited fleet of small rockets and nanosatellites envisioned under the JFC-demand-focused ORS incremental strategy.

This small discussion suggests that "supporting the warfighter" by giving him exactly (and only) what he wants may be counterproductive to military space innovation. Letting the JFC dictate development terms may harm him by not allowing others to "properly teach him to want new things." In order to foster a culture of real innovation, perhaps it is time to reintroduce the services (with their land, air, sea, and amphibious "producer" focuses) to requirements decision making so that environmental "innovation for innovation's sake" can again seek truly valuable innovation in military force development.

Space Power Organization: Aligning Logic and Grammar

Although treated in separate chapters and seemingly concerned with differing attributes of organization, space power logic and grammar are not independent and wholly separate entities. At first glance it would appear that grammar only develops tools and logic only influences the use of those tools. Thus, a "requirements pull" advocate may maintain that logic must take precedence over and dictate the tools the Grammar of Space Power is to develop. Alternatively, a "capabilities push" approach may argue that developments in grammar are the primary drivers of Space Power Logic by providing logic new accesses to develop ability, which will allow logic to transform into national power. In reality, logic and grammar are inescapably intertwined. As Admiral William Holland argues:

> Logic suggests that policy directs strategy, which in turn leads to tactics to execute that strategy. These tactical considerations then become the foundation for supporting technologies. The technologies developed lead to acquisition of the equipment necessary to support the tactics. This logic, adopted from business and economic models, is the basis of the Planning Programming and Budget Systems.
> Experience suggests the real paradigm works differently. Organizational knowledge built on an understanding of environment and mission enlarged by study and experience forms the foundation of tactics. From this basis, an understanding of national interests, a sense of the history of conflict, a grasp of the capabilities of potential enemies, and an appreciation of technology all drive tactical opportunities. These in turn establish the designs for development of technologies and future acquisitions. Equipment developed makes possible improved, advanced, or different tactical possibilities. These new tactics in turn allow changes to strategy. Such changes may or may not then be reflected in policy.[82]

Neither Space Power Logic nor Grammar can claim place of precedence in the development of space power. Because of this fact, any organization dedicated to developing space power must be organized to successfully develop both the tools and intelligent use of space power. What does such an organization look like? To answer this question, we must look to history to identify role models suitable for those who would revolutionize American space development. In Chapter 4, we will explore how a visionary group of officers and leaders successfully organized American sea power for revolutionary development prior to World War II, with dramatic consequences for the United States.

Chapter 4

The Navalists' War—
The Pacific 1941–1945

Because the General Theory of Space Power takes its main inspiration from Mahan's work on sea power, it is fitting that perhaps the best historical example with which to explore how innovation has been generated in military history is from an event sparked and inspired, in part, by Mahan's work. In order to better understand how to stimulate innovation in space power, we must look to see how the U.S. Navy orchestrated the greatest transformation in military history to win the greatest sea war in human history.

Sea Power Revolution

In the late 1880s, Rear Admiral Stephen B. Luce, the leader of the "navalists," like-minded naval officers and civilian government leaders dedicated to an American naval revival, looked at the U.S. Navy as a hollow shell of its former glory and was one of the only officers who could see what it would become. The great fleet that strangled the Southern Confederacy was gone, decomposed into dust in berths scattered across the Atlantic Coast. A mere 92 ships remained of Lincoln's Navy—of those only 32 in commission—and little more than 8,000 officers and men in 1884. By contrast, the Great Britain's Royal Navy possessed 359 ships in commission and almost 64,000 officers and men.[1]

Worse, perhaps, than the number of ships was their overall condition. The fleet was a disgrace to a major nation: wooden hulls, sails, and the most advanced with only rudimentary coal-powered steam engines armed with foreign weapons because no American company could produce modern naval ordnance. Among the world's great navies, America's was not even the best in the New World. Modern steel hulls, rifled shells, and oil-burning engines were unknown in the U.S. Navy.

Few could imagine that only a little more than a half-century later, this broken fleet of fouled bottoms and obsolete arms would be transformed into the most powerful force ever afloat, with thousands of warships and support vessels consisting of massive aircraft carriers, battleships, submarines, fast oilers, and floating dry docks. Furthermore, this fleet would fight the most complex naval and logistical war in human history and win against an empire located half a world away across thousands of miles of water.

The transformation of the U.S. Navy from derelict to dominant did not happen overnight, and certainly did not happen by accident. It required a dedicated and nonstop effort to forge this mighty instrument of national policy, an effort beginning in the 1880s. As historian Mark Russell Shulman notes:

> A new aggressive American naval strategy emerged in the 1880s and nineties as the product of a distinct political agenda formulated and effected by a small group of energetic, progressive, intellectual timocrats. Although the navalists provided the catalyst for the new navy, the process of its creation required popular support. The general public, as well as the political and intellectual elites, determined the shape and consequently the strategy of the new navy. Together, they created an imperial service.
>
> The nation's first peacetime buildup was executed with remarkable speed and thoroughness, due mainly to the extraordinary wide-ranging efforts of the navalists. These men knew that to effect a new grand strategy they had to generate support for a new political discourse, which in turn would engender the strategy. Although it changed through the years, the navalist strategic vision called essentially for a fleet-in-being composed of first-rate ships combined to implement American policies abroad through a concerted force capable of dealing a lethal blow to the enemy on the high seas. This military strategy accompanied a less clearly defined grand strategy in which the United States would claim its place among the great, or even greatest, powers.[2]

The navalist naval strategy made the United States the greatest power on the sea. Navalists had, over fifty years, revolutionized the Navy in every way imaginable: tactics, techniques, matériel. Navalists performed superbly. But it almost wasn't enough.

The Pacific War

The naval war against Japan in World War II, 1941–1945, was the war that the navalists were preparing for. The enormous scale, the vast distances, the merciless climates, the formidable nature of the enemy, every aspect of the Pacific War required revolutions in every facet of the Logic and Grammar of Sea Power. Hundreds of ships, thousands of aircraft, tens of thousands of men, and billions of dollars were spent in the fighting between the Allies and the Japanese Empire. The Pacific War was the purest naval war ever fought, and was indeed the culmination of the navalist project. Without the navalists and their innovations, the United States could not have fought the Pacific War. In order to generate the necessary advances to win such a new and foreign war, the navalists reorganized the U.S. Navy to foment innovation along every line of advance.

Innovation in the U.S. Navy was generated primarily from two thrusts: bureaucratic reorganization, including creating new agencies to support and expand operational and technical innovation; and an extensive program of war games and exercises with which to manage, test, and incorporate these advances. Through the half-century between Luce's dilapidated Navy to the fleet that forced its way into Tokyo Bay in 1945, the U.S. Navy pioneered aircraft carriers, massive battleships, sleek attack submarines, floating bases, and auxiliary craft never before seen, all using a brand new oil-powered propulsion system. In the span of a single lifetime, the Navy was completely remade, and not a moment too soon. In retrospect, the extraordinary efforts of the navalists and their descendants between the 1880s and 1940s weren't merely lucky or useful—they were required. And they required every bit of time and effort available to succeed. But how did they do it? The Navy laid the foundation for innovation by reorganizing the Navy, and instituting a system to test new concepts and technology through a robust program of war games and free exercises. Both provided the needed catalyst for naval transformation.

Organizing the Navy for Sea Power

In the late nineteenth century, and continuing well into the twentieth, the primary administrative units of the U.S. Navy were the bureaus. Each was independently commanded by a flag officer and responsible for a specific part of Navy activities, answerable only to the Secretary of the Navy. The Bureau of Navigation commanded Navy navigation research as well as managing personnel and assignments. The Bureau of Ordnance managed procurement of navy weapons and ammunition. The Bureau of Construction & Repair acquired and maintained the Navy's ships. Other bureaus included Engineering, Yards & Docks, Provisions & Recruiting, Equipment & Recruiting, and Medicine & Surgery. Another major bureau, Aeronautics, was created in 1921.

These bureaus can be considered almost the ultimate in organizational stovepipes. They were very close to being the exact opposite of the horizontal "flat" organizational structures favored by modern specialists to stimulate innovation. Yet, the bureau system lasted from 1842 to 1966, throughout the navalists' transformation era. The navalists didn't attack the bureau system or see it as a detriment to transformation. Instead, they advocated for three new organizations that existed apart from, and above, the bureaus and led by the Secretary of the Navy, to foster cooperation and communication across the Navy enterprise. These three organizations were the Naval War College, the Navy General Board, and the Office of the Chief of Naval Operations.

Historian and naval officer Commander John Kuehn describes the interplay between these three agencies during the period between World War I and World War II:

> The Navy of the interwar period was a collaborative place and the General Board encouraged cooperation, and thus innovation, across the various levels of war—tactical, operational, and strategic. The Board tended to focus most on the strategic level. It developed policy and applied its policy decisions to the overall design of the Navy (numbers and types of ships). OpNav [Chief of Naval Operations or CNO]—inherent in its name—was the operational level of war that included planning for both current and future operations. Although its War Plans Division included a strategy "cell," OpNav focused overwhelmingly on the operational spectrum—conducting real world operations and exercises and designing plans around the fleet at hand. OpNav collaborated with the Naval War College in testing and developing operational concepts using the College's war gaming process. The results of these war games were also shared with the General Board. Similarly, the War College was also linked to the General Board by statutory membership. Tactical issues came up in the course of planning at OpNav, in war gaming at the War College, and in the discussions of the General Board. At the Board's hearings and meetings, technical and tactical design considerations entered the process, often through testimony by representatives of the bureaus and other experts. These meetings, especially in the topical hearings, became a forum for the discussion of policy and design issues at the strategic, tactical, and operational levels of war (see Figure 4.1).[3]

The interconnection between the General Board, Naval War College, and Chief of Naval Operations is critical to understanding how the Navy was able to transform itself in time to fight the Pacific War. In order to best understand why, we must explore each of these organizations individually.

THE GENERAL BOARD

The General Board of the United States Navy was established on 13 March 1900 to act as the first naval general staff and lasted until 1951. It

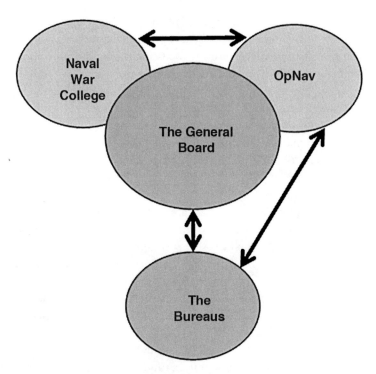

Figure 4.1. U.S. Navy Organizational Relationships during the Interwar Period (Kuehn, 13). Overlap indicated membership and lines indicate communication and coordination.

emerged from lessons learned during the first war encountered by the navalists, the Spanish-American War in 1898. J.A.S. Grenville explains:

> The war with Spain had underlined the need for adequate staff work and the success of the War Board had pointed the way for the future. Among the most persistent advocates of a general staff for the Navy was Captain H. C. Taylor. He had first laid plans for such a staff before Roosevelt in May 1897; now in 1900 he brought the idea once more to the attention of Secretary Long. Long, however, was reluctant to risk a fight with his entrenched bureau chiefs, hesitant about allowing the professional officers wide powers outside civilian control, and rightly dubious whether Congress could be brought to approve the scheme. Consequently he compromised, and in March 1900 created a Board, known as the General Board, which possessed no executive functions, but was to serve as a purely advisory council which was constitutionally confined to considering such problems of strategy as the Secretary of the Navy might refer to it.[4]

The General Board's membership alternated between seven and fourteen members throughout its life, but consisted of the executive committee and ex-officio members. The executive committee often consisted of senior officers of the Navy and Marine Corps near the end of their careers with no line responsibilities. They were charged with deliberating and all manner of issues affecting the Navy with a dispassionate eye and focused entirely on problem solving. Ex-officio members of the board included officers appointed to important positions in the Navy, including the chief of the Bureau of Navigation, the Director of Naval Intelligence, and the President of the Naval War College. Together, this group of senior officers acted as senior advisors to the Navy Secretary.

Kuehn argues that the General Board was far more than a merely advisory body. "The General Board is often described as merely an 'advisory body' to the Secretary of the Navy, however, its real functions covered all matters of policy and strategy pertinent to the Navy from 1900 to 1950. This unique organization influenced innovation because of its balanced membership and its organizational function as the nexus where policy was translated into force structure."[5] In fact, Kuehn concludes that "the General Board played the critical organizational role in linking the [limitations of the Washington Naval] treaty system with innovation in the design of the fleet."[6]

In effect, the General Board acted as a clearinghouse for ideas that sprung from the independent bureaus and the Navy as a whole and were discussed in a frank and open manner. There the opportunities of new technology such as naval aviation strike craft, developed and promoted by the Bureau of Aeronautics, could be discussed with senior officers and other bureau representatives from Navigation and Engineering, resulting in common understandings and agreement on future use. Indeed, the General Board served as a way for new ideas to be considered, refined, and incorporated into (or discarded from) the fleet. The General Board could then be used as the forum for accepting new ideas (transformers and combinations, as defined by the General Theory of Space Power) and incorporating innovation into the Navy as a whole at the strategic level. However, generating and developing new transformers and combinations for sea power would require a different organization.

THE NAVAL WAR COLLEGE

The Naval War College opened its doors on 6 October 1884. Its founder and chief proponent, Admiral Stephen B. Luce, speaking of the intent of the college, declared:

> It is a place of original research on all questions relating to war and to statesmanship connected with war, or the prevention of war. That "war is the best school war," is one of those dangerous and delusive sayings that contain just enough truth the secure currency: he who waits for war to learn his profession often acquired his knowledge at a frightful cost of human life.[7]

Accordingly, the Naval War College played the role of idea generator, first tester, and trainer of new ideas to commanders of the fleet. The college accomplished these missions primarily through the use of the naval war game, a table-top system of rules (the antecedent to modern commercial and military war games alike) that allowed students and instructors to simulate naval war on grand strategic, operational, and tactical scales from the comfort of a classroom.

The importance of the Naval War College war game to the development of the officers who would fight the Pacific War cannot be overstated. Fleet Admiral Chester Nimitz said in 1965, "The enemy of our games [at the Naval War College] was always—Japan—and the courses were so thorough that after the start of World War II—nothing that happened in the Pacific was strange or unexpected. Each student was required to plan logistic support for an advance across the Pacific—and we were well prepared for the fantastic logistic efforts required to support the operations of the war. The need for mobile replenishment at sea was foreseen—and practiced by me in 1937."[8] Great praise, indeed. But naval analyst and war gamer Peter Perla goes much farther:

> The archives of the Naval War College have preserved the records of more than 300 war games played during the interwar period. Of these, more than 130 emphasized campaign-level or strategic play, with all but nine of those focused on a possible war with Japan. During the course of these games, the discussions and critiques of players and staff evidenced a growing appreciation for the strategic realities that would later surface during the war in the Pacific. From an initial focus on a Mahanian vision of an early, decisive clash of battle fleets, which would decide the outcome of the war in an afternoon, the games evolved into a more realistic and grimmer version of a prolonged struggle, not just between fleets, but between nations and societies. During the process, the young officers who would rise to the command of task forces and fleets in the coming war saw their perceptions torn apart and then rebuilt.
>
> Much has been made of the famous statement of Admiral Nimitz that nothing that happened in the war was a surprise except the kamikazes. More interesting, and probably more indicative of the value of the games in the development of Nimitz's and the navy's strategic thinking, are the words he wrote in his 1923 thesis: "To bring such a war to a successful conclusion Blue must either destroy Orange military and naval forces o effect a complete isolation of Orange country by cutting all communications with the outside world. It is quite possible that Orange resistance will cease when isolation is complete and before steps to reduce military strength on Orange soil are necessary. In either case the operations imposed upon Blue will require the Blue Fleet to advance westward with an enormous train, in order to be prepared to seize and establish bases en route.... The possession by Orange of numerous bases in the western Pacific will give her fleet a maximum of mobility while the lack of such bases imposes upon Blue the necessity of refueling en route at sea, or of seizing a base from Orange for this purpose, in order to maintain at least a limited degree of mobility."

Over the course of the interwar strategic games the type of careful and perceptive thought exhibited by Nimitz reshaped the way the navy came to think of its role in a future war with Japan. One historian of the War College games has argued that "the repeated strategic gaming of Orange war forced the Navy to divest itself of several former "reality-assumptions":

- The notion that war at sea was defined according to a formal, climactic clash of battle fleets, and that naval strategy consisted of maneuvering one's fleet to bring the adversary to decisive engagement.
- The belief that superior peacetime naval order of battle was equivalent to available force in war, that a peacetime status quo would persist indefinitely, and that only traditional naval weapons according to traditional hierarchies of importance would be necessary to defeat the enemy.
- The assumption that naval war across an oceanic theater could be conducted quickly, and that enemy advantage in strategic geography was marginal both to strategic planning and to the conduct of naval operations in war.
- The hypothesis that war with Japan would be limited in forces engaged, in objective, in belligerent participants, and in time."

By emphasizing the broad questions of strategy and managing a fleet at the expense of downplaying some of the technical and tactical details of combat, the War College's strategic games left the players free to explore their options and to teach themselves the deeper truths of the war that was coming. Based on the insights afforded by strategic gaming, the navy began to explore the requirements for a measured, step-wise offensive campaign to span the Pacific, requirements not just for the navy, but for the entire national political and military apparatus. "Gaming reality forced the Navy to seize a set of strategic concepts about the conduct of future war which had the capacity to redefine the very nature of America's role in the world."[9]

The War College, in its exploration and training role, generated new ideas for the Navy, tested them through the inexpensive and versatile war gaming process, and inculcated lessons learned about new technologies and operational concepts to the commanders of the Pacific War. The impact of the Naval War College on the Pacific War can be summed up:

By the time Japanese forces attacked Pearl Harbor in December 1941, the Naval War College had already made its most significant contribution to the war effort by its earlier training of officers in a methodology for problem solving.... When the United States entered World War II, every flag officer qualified to command at sea, but one, was a graduate of the Naval War College, and had become accustomed to think in terms it had established.[10]

The Chief of Naval Operations

The final critical navalist organizational innovation was instituted on 11 May 1915. The Office of the Chief of Naval Operations was the planning element of the Navy for current and future operations, and it was in its planning and exercising mission that "OpNav" made its largest impact on innovation. Henry Beers describes the Navy's organization in May 1916 from a statement by the Chief of Naval Operations:

The manner in which the Office of the Chief of Naval Operations was working by the year 1916 is shown in the following statement by its chief:

Section I—Information. Before any intelligent plans can be made or any effective work accomplished, complete and comprehensive information is necessary. The duty of collecting and disseminating information is assigned to the Office of Naval Intelligence.

Section II—Education. This section is represented in the main by the Naval War College, where the conduct of war on the sea is made the special object of research and study, and where officers are given the opportunity to train themselves in the conduct of war.

Section III—Planning. This is the deliberative section to which information gathered by Section I and experience and training gathered by Section II are brought. Here, free from all administrative work, plans are deliberated upon, perfected and submitted for the Secretary's approval and final adoption. This section is now the General Board, of which the Chief of Naval Operations is a member. It fills a vital office of the general function of preparing the fleet and operating it in war.

Section IV—Inspections. This section applies the test of inspection. It tests the preparedness of the fleet for war. Material preparedness of the ships is tested by the Inspection Board, gunnery preparedness by target practice, and steaming efficiency by annual steaming competition.

In that connection I might state that testing the preparedness of the fleet for war is accomplished or is expected to be accomplished—it has been begun by the so-called maneuvers. We have had two during my term of office, one that was formulated before I took charge of the office and another after I took charge of the office. The method in which these maneuvers are performed is intended to give the department a clear and definite idea as to the general preparedness of the fleet to meet an enemy.

Section V—Execution. This is the last section. It is charged with the duty of carrying out approved policies and plans, and is under the immediate supervision of the Chief of Naval Operations.

The organization which carried out the foregoing general plan was as follows:

1. Division of Operations—organization of the Office of Operations; naval policies; military characteristics of ships, building programs; fleet maneuvers; and organization and employment of fleets.
2. Plans—plans, war portfolios, and reports on preparedness for war.
3. Naval districts—operations of mining and operations of naval districts.
4. Regulations, records—aid to Chief of Naval Operations, Navy Regulations, general orders, custody of files, and matters requiring coordination of other executive departments.

5. Movement of ships—movement orders—that is, military movements; schedule of employment of forces; records of service of fleets and vessels; military operations of naval forces on shore and reports of naval operations; movement orders (logistic); logistic establishments afloat, and naval auxiliary service.

6. Communications—and that includes the radio, cable, telegraph, and telephone service—censorship, signals, and codes. Communications office.

7. Publicity—that is, general news and movements given out to the public, distribution of current communications, censorship, and press notices.

8. Matériel division—logistics of the office; location, characteristics, and development of bases, supplies, and reservations, and coordination of material bureaus.

Such was the outline of the organization of the Office of the Chief of Naval Operations as it existed in March, 1916.[11]

This order not only better described the goals and divisions of OpNav, but also explains those of the Naval War College and the General Board with which it helped shape the U.S. Navy into the navalists' vision in time to fight the Pacific War. OpNav had a hand in almost every activity in which the Navy was involved. The key in its effectiveness is that OpNav focused on the operational level of war, and it is at the operational level of war where new technology and tactical ideas meet the needs of national strategy to find satisfactory answers to necessary questions. At this operational nexus OpNav placed perhaps the largest and complex scheme of military maneuvers ever devised, the annual U.S. Navy fleet problems.

Training the Fleet for Navalist War

Albert Nofi, the author of the definitive account of the interwar Navy fleet problems *To Train the Fleet for War*, describes a key factor in America's defeat of Japan in World War II:

Between 1922 and 1940, the U.S. Navy conducted an extraordinary series of major free maneuvers called "fleet problems." Writing in 1939, Secretary of the Navy Claude A. Swanson noted that the fleet problems were "of the utmost value in training the personnel of the fleet." The fleet problems were an attempt to engage in maneuvers under conditions that approximated those that would occur in actual war. Decades later, Admiral James O. Richardson, who had served as Commander-in-Chief United States Fleet (CINCU.S.) in 1940–41, recalled that the fleet problems were "fought with zest and determination" limited only by considerations of safety.

It was during the fleet problems that the U.S. Navy evolved from a force that thought of the future in terms of a series of somewhat more sophisticated battleship clashes in the style of Jutland, to one that saw the future, albeit unclearly, in terms of surface, air, undersea, and marine forced integrated into a combined arms "naval force" capable of carrying American power across the Pacific to Japan.[12]

It was in the major exercises known as the fleet problems that the Navy provided the innovational grist for their reorganization to forge into a mighty weapon. The Naval War College–General Board–OpNav triad could provide the structure with which to invent, develop, debate, and decide upon innovative ideas, but it was the fleet problems that gave the planners feedback on effectiveness and the all-important real world experience required to make an initial idea an understood and effective reality. It was in the fleet problems that the Navy brought many concepts to fruition, like naval aviation doctrine and advanced logistical concepts, and discarded others, like airships. While some ideas could be simulated on paper in the Naval War Game, only in real world operating conditions under simulated warfare could senior officers understand these concepts' best wartime use and train junior personnel how to perform these wartime duties effectively.

In fact, the fleet problems connected the two major revolutions in naval warfare that occurred after the rise of the navalists in the late 19th century. When the push for naval reform came, steel-hulled battleship children of *Dreadnought*, submarines, and torpedo destroyers that fought World War I were mere concepts with no real world experience upon which to build. The Jutland-style naval battle was in itself a revolution in naval warfare. Therefore, the fleet problem in the beginning was focused on fighting a naval war that was remarkably modern. The institutional Navy had no real reason to expect that another massive change in fighting was to be required. So how did the fleet problems exchange one modern and innovative Navy for another?

The fleet problems were not exercises like those conducted by the American military today: scripted, predetermined scenarios more interested in ensuring commanders and units can conduct their assigned missions rather than exploring how things may develop through the fog of war. The fleet problems were "free maneuvers," meaning that the Navy pitted one fleet against another, gave very broad orders, and allowed the fleet engagements to evolve as they may. Nofi explains:

> The fleet problems were genuine free maneuvers, involving most of the available resources of the entire Navy, portions of which operated against each other under the leadership of the officers who would command it in wartime in highly unscripted campaigns that sometimes sprawled over hundreds of thousands of square miles of ocean, which provided valuable lessons for the development of the Navy, many of which are still relevant.[13]

The wide latitude given the fleet problems' two opposing commanders ensured that the fleet problems would be used as a test bed for experimentation in nearly all areas of naval endeavor. The fleets would have to sail, meet their objectives, and frustrate the enemy objectives, with whatever forces were available. Tactical exercises, like gunnery and logistical operations, were in the problems but they were not the extent of the problems. Because of this free hand, innovation was encouraged across the fleet. The fleet responded accordingly. Over twenty years:

> the fleet problems affected the Navy on many levels. The most important of these was that they helped train the fleet, and particularly its commanders and their staffs, in conducting protracted transoceanic naval campaigns, and they helped develop the carrier task force and carrier air doctrine. But the other contributions, from tactical development to technological experimentation to public relations, all played a role in the development of the Navy during the interwar period, and contributed to the crafting and refinement of War Plan Orange as the service prepared itself for a trans–Pacific war with Japan.[14]

The fleet problems, however, weren't entirely unscripted. The fleet problems were intended to train the fleet for the wars they would most likely be called upon to fight. And after 1898, with the annexation of the Philippines, the leaders of the United States Navy believed that the war they would most have to prepare for was one against Japan. This was not necessarily because they believed war with Japan was inevitable or desirable. The defense of the Philippines, so far from the United States, simply had the combination of possibility and challenge to make a most interesting Pacific War scenario. Simulating this war alone drove innovation because of the many techniques and operations that would be required in fighting across such large distances. Nofi describes the geopolitical factors in the fleet problems:

> Though it may at times not be readily apparent from the documents, the fleet problems primarily focused on only one strategic concern, war with Japan. The specific problems devised for the fleet were almost always related to that overarching strategic question. Fleet problems were usually intended to explore the fleet's ability as a fleet to execute some aspect of, or contingency related to, War Plan Orange. This was the case even when a European power was vaguely identified as the "enemy."
> Of course War Plan Orange envisioned operations across the western Pacific, and it was unrealistic—as well as diplomatically impossible—to conduct maneuvers in the actual potential theater of operations. As a result, during most fleet problems there was a great deal of what CINCUS Robert E. Coontz termed "geographic transposition and orientation." That is, the geography of a sizable chunk of the globe was notionally rearranged to permit the fleet to operate in an environment that most closely resembled the probable theater of operations. The actual kinds

of operations that were to be included in the fleet problems flowed naturally from the need to conduct operations across such vast areas. So it was routine to include exercises in fleet escort and defense, scouting and evasion, underway refueling, and opposed entry into a friendly port, as well as battle line tactics, fleet submarine operations, air warfare and fleet air defense, and landing operations, in scenarios that dealt with different possible aspects of the expected war. At times the scenario for a fleet problem, or a particular phase of a problem was inspired by something that had occurred in an earlier problem. There was systematic interaction between the fleet problems and wargaming at the Naval War College, a process that resulted at least in part from the introduction of better record-keeping procedures for wargames in Newport, beginning in 1922. Thus, ideas developed or problems encountered on the game floor were often examined during the fleet problems and vice versa. On at least one occasion, during joint army-navy coast defense maneuvers that followed Fleet Problem VII (1927), students at the Naval War College played the same scenario on the game floor that the naval, air, and ground forces were playing in Rhode Island Sound and adjacent coastal areas, which seemed to have enriched the subsequent critique of the problem.[15]

Development of the fleet problems was, by necessity and design, a cooperative effort across the Navy as a whole. Although the Office of the Chief of Naval Operations was in direct control of the exercises, problem development was never solely an OpNav endeavor. Indeed, the annual fleet problem was one of the major tools used by the Naval War College–General Board–OpNav triad to drive innovation, and the problem helped cause that innovative organizational structure to succeed. Planning for the fleet problem was always took great effort:

> Once the CNO and CINCU.S. had reached agreement on the nature of a [fleet] problem, actual planning began. Planning for a problem was done by officers on the CINCU.S. staff, often assisted by personnel from the Naval War College. They were at the apex of a complex process in which the commanders and staffs of the principal components of the fleet were solicited for comments on the proposed plans and were free to make suggestions about operational and tactical ideas. These comments and proposals were studied by the CINCU.S. staff for possible incorporation into the final plans.
>
> This helped provide a realistic degree of continuity combined with a healthy dose of innovation, by ensuring input from the service's senior-most officers, the General Board, and the Naval War College—who collectively were both the preservers of the Navy's "institutional memory" and responsible for its future development—as well as from the actual commanders who would have to conduct operations if war came.[16]

The fleet problems gave the new organizational structure of the Navy increased meaning and effectiveness, and every bit of efficiency was desperately needed. The Navy had plenty of problems that needed to be worked out. It was dealing with a great amount of new and untested technology, new operational requirements to be able to fight a major war halfway across the world, and had a very limited budget with which to work. The fleet problems, being as devoted to open experimentation and the "settling" of major problems being discussed by the Navy triad, were ideally suited for solving the Navy's problems:

> By providing the service with an "instrument" that permitted maneuvers to unfold in as realistic and unrestricted a way as was possible short of actual war, the fleet problems served not only as a test of the Navy's skills, but also as a laboratory in which to test existing and new warfighting ideas and technologies. The fleet problems helped the Navy not only to improve its mastery of the tools of sea power, but also to sort out what worked and what did not from a plethora of new ideas, technologies, and capabilities, while giving commanders maximum opportunity to formulate creative solutions to realistic situations. This process created the naval force that secured for the United States command of the seas in the Second World War and the Cold War—and into a new century.[17]

Concepts, technologies, capabilities, command flexibility. All needed to be tested and refined in order for the Navy to be ready for the Pacific War. The easiest changes to recognize were those in new technology and warships, but as Nofi indicates, new technology was useless unless people knew how to fight with it. Those in charge of the fleet problems were aware of this:

> The fleet problems enabled the Navy to test a variety of new ideas and equipment under realistic operating conditions. It was during various fleet problems that a number of innovative doctrinal concepts were first tested under "wartime" conditions, such as the carrier task force, underway replenishment, combat air patrol, and circular cruising formations....[18]
>
> Although most references to the fleet problems focus on their importance in the evolution of carrier aviation, they were in fact critical to developments in every warfighting area and in every aspect of technology and the operation and management of ships, aircraft, and personnel. There were always several experimental threads running through a fleet problem. These thread usually recurred year after year, each time being examined in a different way, due to increasing experience, better technology, new ideas, and changing plans.
>
> As doctrinal ideas were refined they were published in the Fleet Tactical Publications, or disseminated by means of lectures at the Naval War College and the annual report of the Commander-in-Chief of the United States Fleet. This kept the fleet current on changes in doctrine.[19]

Incorporating new technology was a hallmark of the fleet problems. The most famous naval technology to emerge after World War I, and most critical to the Pacific War, was the aircraft carrier. Popular imagination believes that after the American battleships were sunk at Pearl Harbor in 1941, the U.S. Navy was forced to rely on its untested and untrained aircraft carriers to keep Japan at bay—essentially forcing the Navy to shift its focus from battleships to carriers. While there is some truth to this statement, the fleet problems refined the Navy's ability to hone the independent aircraft carrier as a formidable weapon well before Pearl Harbor. James M. Grimes clearly summarized the importance of the fleet problems to the evolution of naval aviation by noting that "it was in the War Games and Exercises of the 1920s and 1930s that Naval Aviation got its practical training and experience through which it developed to a point where it was enabled to prove its worth and to take its place as an integral part of the fleet."[20]

During the fleet problems, the aircraft carrier evolved from an auxiliary scouting tool into a major capital ship. The emergence of the fast aircraft carrier task force was not a sudden innovation forced upon the Navy in order to fight the battle of Midway, but was already a common fleet formation practiced for years in the fleet problems. While aircraft carrier warfare was perfected during the Pacific War, it was placed on a firm foundation well before the aircraft was used in anger, due to the fleet problems.

However, the aircraft carrier was not the only innovation that emerged from the fleet problems. Aircraft carriers extended the fleet's striking ability to hundreds of miles, but the Pacific was still orders of magnitude larger. Fleets would have to move thousands of miles before they even got close to the combat zone. In the beginning of War Plan Orange planning, coal was still used as the Navy's primary fuel source. The emergence of oil in the early twentieth century made traveling large distances much more economical, but it alone didn't solve the crushing tyranny of distance. Once again, the fleet problems operationalized the solution used in the Pacific War:

> The perfection of underway refueling by the "riding abeam" or "broadside" method, first developed during World War I and still in use today, was the most notable innovation in fleet sustainment to emerge during the fleet problems. This method of refueling of smaller warships introduced during Fleet Problem II (1924) because "almost a standard part of each fleet problem during the late 1920s and early 1930s." By Fleet Problem XIII (1932), experiments in the simultaneous refueling of two vessels from a single oiler were undertaken, and this practice also shortly became routine, as did the refueling of destroyers from larger warships. Nevertheless, despite proposals as early as Fleet Problem II to attempt refueling of heavier ships by this method, more than a decade was required to overcome technical and cultural barriers, ultimately requiring the intervention of the CNO. Despite these obstacles, the refueling of battleships and carriers by the riding abeam method was successfully demonstrated during Fleet Problem XX (1939), and also shortly became a matter of routine.
>
> The fleet problems also helped convince naval leaders of the importance of procuring high speed auxiliaries, oilers, troop transports, and cargo vessels. Recommendations to procure auxiliaries able to sustain 12 knots or more were made in the final report for every fleet problem from the very first, in 1923. Moreover,

the development of War Plan Orange indicated the need for literally hundreds of auxiliaries, oilers, troop transports, and cargo vessels, as well as many specialized vessels. During the 1920s and the early years of the Depression, however, neither the public nor the Congress were inclined to spend any money on the Navy. Even when funds began to become available during the Roosevelt administration, the focus was on building the fleet up to allowable treaty limits. Given that it took longer to build warships than auxiliaries, this was probably a good decision, particularly given that the Merchant Marine Act of 1936 began an important reorganization, standardization, and expansion of the country's merchant fleet. Nevertheless, on the eve of Pearl Harbor the Navy did not have enough modern fleet auxiliaries. The resulting shortage of high speed auxiliaries, and notably fast oilers, hampered operations during the first two years of World War II.[21]

The fleet problems pointed the way to solving the distance problem inherent in fighting the Pacific War: underway refueling and high-speed auxiliaries. Fleet sustainment was perhaps an even greater burden to fighting the Pacific War than characterizing carrier warfare. One can envision a Pacific War fought between battleships, but it is hard to imagine one without fleet oilers. Even so, this importance is deceptive because it is such a mundane problem. Logistics is simply not as sexy as air combat. But despite this lack of excitement and appeal, the fleet problems caused advanced maritime logistics to be imprinted upon the American naval mind:

> The evolution of underway refueling presents a good example of how an innovative technology worked its way into the Navy's tool box. During the early fleet problems underway refueling by the riding abeam method clearly was a highly experimental procedure.
> Within a few years underway refueling became "almost a standard part of each Fleet Problem." By the later fleet problems it was becoming common for larger warships to be refueled in this way as well, despite some trepidation on the part of a few senior officers.[22]

Without this extensive practice of underway refueling the fleet problems provided, learning or accepting these techniques while fighting the Japanese would have proven exceedingly difficult, if not impossible, for the U.S. Navy to internalize. If it would have at all, the cost in blood and treasure would have been remarkable.

Many technological advances were accepted and internalized during the fleet problems, but not all technologies tested were found appealing. Despite large interest, especially among naval aviators, the Navy turned away from the great airships during the interwar years, primarily due to their poor performance during the fleet problems. Nofi notes the experience of the airships in the fleet problems:

Fleet Problem XIII added further evidence that the airship was a weapon of doubtful value. Proponents of zeppelins tried to brush aside the poor performance of *Los Angeles* (ZR 3) as a result of her age, having been built in 1924, while touting the merits of the new "flying aircraft carriers *Akron* (ZRS 4) and her sister *Macon* (ZRS 5), neither of which had taken part in the maneuvers. Nevertheless, in a prescient comment to CNO William V. Pratt, Admiral Schofield said, "the need for aircraft is not more Akrons, but more carriers," which seemed to have summed up the fleet's opinion well.[23]

Thus, the fleet problems allowed some technology into the U.S. Navy, and rejected widespread deployment of others. Acting as a remarkable discriminator of ideas and technologies, the fleet problems certified innovation in the Navy. This is just one way that the fleet problems helped spur innovation in the Navy. Because of the fleet problems' success:

> It is hard to argue with the conclusion drawn by Admiral James O. Richardson that the fleet problems "were expensive in time, money, and effort, but they led to great advances in strategic and tactical thinking which marked our naval development during" the interwar period. As Secretary Claude A. Swanson had observed in 1939, the fleet problems were "of the utmost value in training the personnel of the fleet."[24]

But how and why did the fleet problems encounter so much success in its almost twenty-year run? There are a number of reasons the fleet problems became such a remarkable tool for the Navy. Perhaps the most important reason the fleet problems succeeded was due to the way leadership approached them. Historian Nofi posits, "The high degree of open-mindedness and flexibility displayed by many senior officers during the fleet problems was quite remarkable."[25] Indeed, where often the military mind is ridiculed for being close-minded and unreceptive to new ideas, the fleet problems seem to exemplify the exact opposite. Without this flexibility across the scope of the fleet problems' leadership, from the most senior fleet commander to the most junior ship division officer, it is difficult to imagine the fleet problems incorporating so many new ideas in service to transforming the way the Navy fights. However, there were many other ways the fleet problems were unique.

Also vastly important to the success of the fleet problems is how seriously the fleet problems were taken after the operational portion of the fleet problem was over. Nofi explains:

> Normally, each fleet problem was followed by a critique, of which James O. Richardson (CINCUS, 1940–41) later wrote, "The battles of the Fleet Problems were vigorously refought from the speaker's platforms." These critiques were often long.[26]

The fleet problem critiques, which often lasted for days or weeks immediately after the end of the problem and were open to all participants and observers regardless of rank, were critical for the Navy's development. Lessons learned from operations are only useful if they are understood and internalized. In order to be understood, they must first be made apparent. By refighting the battles via the speaker's platform and hearing the motivations, experiences, thought process, debates, and motivations of each side, all participants could better understand why events developed as they did. Without this ability to see all available information, the wrong lessons could easily be learned if a major factor was overlooked. The immediacy of the critiques allowed for reflection with everything fresh in participants' minds, but that just leads to the correct lessons (as much as possible) being recorded. It was the openness of the critiques that brought these lessons to the fleet:

> The conclusions, which were often frank, the lessons learned, and the recommendations usually incorporated excerpts of the individual reports, and at time entire individual reports, of the principle subordinate commanders. This was done even when the reports of these subordinate commanders did not agree with the conclusions and recommendations made by CINCUS. or the two fleet commanders. The comprehensive fleet problem report was published and widely circulated for study, evaluation, and comment. At times, developments in a fleet problem sparked experiments in wargames at the Naval War College. These experiments would explore ideas that might be tested in a later fleet problem, as part of a "simulation–exercise (and fleet problem)–simulation cycle."[27]

Not only were the lessons explored and written down, they were used by the institutional Navy. The publication and wide distribution of the fleet problem report, complete with dissenting viewpoints, ensured that the information was available to all Navy personnel interested in fleet problem events or with an interest in a particular segment of the problem, such as aviators examining the results of fleet air operations. However, these lessons were also used as feedback into the critical simulation–fleet problem–simulation cycle that took advantage of the critical Navy organizational triad to maximize fleet innovation and development. The honest, timely, and complete critiques of the fleet problems were instrumental in the fleet problems' performance in revolutionizing the Navy.

Of course, an obvious, but overlooked, unrecognized, and vastly important reason the fleet problems were successful is that the fleet was available to devote to experimentation:

> From about 1922 until well into the later 1930s, the bulk of the fleet was hardly ever called upon to show the flag, and it never had to fight. As a result, as naval analyst Peter M. Swartz noted, the fleet became "a giant training center and laboratory, and its operations giant training drills and fleet battle experiments." In the fleet problems, "the role of at-sea exercises as not only a schoolhouse but also the

fleet's laboratory for experimentation and testing of innovative ideas reached its apogee."...

The fleet problems were intended to provide the Navy, and particularly its senior commanders, with the most realistic possible training short of actual combat, to teach officers how to think through problems and train them in the development of operational plans and orders, and to test doctrines and technologies. They marked the culmination of the Navy's training year....[28]

In retrospect, having a large military force available to be dedicated for experimentation for long periods of time is a rarity. The modern U.S. military, despite its size, is often stressed to its near breaking point due to operational requirements. To devote any time to simple experimentation not directed toward a critical near-term interest is almost impossible. Fleet problem planners and leaders were aware of the difficulty in conducting effective experimentation during times of high operating tempo:

> Writing after World War II, James O. Richardson observed that preparing the fleet for possible war was incompatible with conducting extensive experiments. During the last years of peace the rapid expansion of the navy and operational requirements detracted from the fleet problems, with the fleet problems arguably detracted from readiness. Moreover, there was little time after a fleet problem for effective reflection upon and analysis of the lessons learned.[29]

Not only must military forces be available for experimentation, there must also be adequate time to assess the results of that experimentation. "Facing no immediate 'threat,' the [interwar Navy] had the leisure to conduct extensive maneuvers and experiments, as these did not interfere with readiness."[30] If time for both is difficult to find during times when readiness for war is essential, it is near impossible during times of actual conflict and active military operations. Nofi addresses this conflict, and issues a warning, by contrasting the activities of a peacetime U.S. Navy in the 1930s with the Imperial Japanese Navy, which was at war with China for much of the same time period:

> Nevertheless, while operational demands may sometimes have limited the value of the fleet problems, the experience of the Imperial [Japanese] Navy was perhaps worse. Beginning in mid–1937, with the outbreak of full-scale war against China, the Imperial Navy seems to have found little time for grand fleet maneuvers. While many useful lessons were undoubtedly learned during active naval operations against China, these operations were all largely in support of ground operations. The net effect was that, although the Imperial Navy's officers and men gained valuable tactical and administrative experience, they were no longer practicing for operational and strategic missions.[31]

Nofi's observation should act as a stern warning to the United States military, especially toward its fledgling and immature space forces: being constantly at war provides great tactical experience, but it dulls operational and strategic development. With the space forces' mantra of "warfighter integration" since 2001, space's contributions to the tactical fight have increased immensely. However, the Air Force Weapons School Space Division's tactical focus only accounts for one-third of the levels of war. Even accounting for space forces' natural impact at the operational level, is readiness and wartime use dulling strategic development? History may indicate that space forces, like the IJN in the 1930s, may be "fighting the last war"—which is the current war—and not devoting adequate attention to the next war.

Money is another important consideration for the fleet problems. The fleet problems occurred during moribund times for defense budgets. Post–World War I "peace dividends" and the Great Depression combined to make money very scarce during the interwar years, and throughout the fleet problems' existence. However,

> Despite the additional funds [provided by the Roosevelt administration], there still did not seem to be enough money for everything that the Navy wanted to do. Nevertheless, although the budgets for most of the interwar years were considered very austere, cost does not seem to have been so important a handicap to the fleet problems as is often claimed in the contemporary literature. Indeed, an argument can be made that that constrained budgetary situation forced the Navy to become more efficient. Moreover, had more money been available during the "austere" years from 1922 through the mid–1930s, much of what those funds would have bought would have been obsolete or obsolescent by the late '30s, particularly in the case of aircraft.[32]

This interesting lesson seems to indicate that in times of rapid technological expansion but limited funding, budgets might better be spent on operational development and training rather than matériel acquisition, for although new weapons might be made quickly obsolete, innovation in tactics, doctrine, and operational understanding has a shelf life that provides a far greater return. Perhaps in today's equally constrained fiscal environment, and the threat of rapid space technological development (such as commercial human space travel, propulsion physics, and exploration programs), the limited money available to develop the space forces should be spent primarily on concept development and technological experimentation rather than fielding large forces for readiness purposes. This may be a subtle, but critical, lesson the fleet problem can provide.

The greatest lesson, perhaps, the fleet problems can provide is the most straightforward: "It required nearly twenty years of fleet problems to develop the organization, tactics, and technologies that enabled the Navy to win World War II."[33] Innovation did not happen overnight. It took the better part of a career in the Navy to see it developed from the battleship-centric fleet of World War I to the aircraft carrier–capable fleet that won World War II. It took years of experimentation, simulation, exercises, and reflection to create the fleet that could win the Pacific War. The new technology, new concepts, and new warships and auxiliaries that the Navy needed to win the war the navalists saw coming took a great deal of time and effort. If we are not willing to devote an equal or greater amount of resources to modern military exploration, we will fail to achieve the same results; results that may be essential in preparing for the next war.

Nofi summarizes the fleet problems as such:

> The fleet problems addressed particular strategic questions identified by the CNO, often drawing upon exercises gained from wargames at the Naval War College or to test plans developed by the [General Board and OpNav's] War Plans Division....
>
> Working in a financially constrained environment, with a mix of old systems, upgraded systems, and new systems, the Navy managed to solve virtually all of the problems inherent in conducting a major maritime war on a global scale, while exploring and developing the basic principles of such fundamental tools of naval warfare as carrier task force operations, amphibious landings, underway refueling, and more, while conducting experiments with new technologies in communications, radar, camouflage, and so forth, thereby creating the fleet that not only won the Second World War, but contributed significantly to victory in the Cold War, and continues to dominate the world's oceans and to project power deep inland.
>
> The fleet problems were the way the U.S. Navy learned to fight World War II, and are an outstanding example of how a military service can educate itself.[34]

To determine how the space forces can and should educate themselves for the tasks they will be asked to accomplish in the future, the interwar U.S. Navy annual fleet problems is a remarkable model to emulate. However, the battle fleet is only one part of the navalist comprehension of sea power. We will now look at another avenue of navalist development.

The Merchant Marine

Just as the General Theory of Space Power states that the military arm is only one point of the Logic Delta, sea power does not consist only of warships and the institutional development of the Navy. The navalists had long been aware of the important connection between a nation's navy and her merchant marine. Admiral Luce expanded on this importance in 1903:

> An intelligent study of naval policy must necessarily include our shipping interests. The military marine and the mercantile marine are interdependent. The navy, while

policing the sea, protects our foreign commerce, and in time of war, finds there its greatest reserves. It was once observed that we had "clipped the wings" of commerce and driven our carrying trade to foreign bottoms. The same is practically true today. Thus we are not only contributing indirectly to the support of foreign navies, which may be some day opposed to our own; but we are depriving ourselves of what would prove, in time of war, an auxiliary of incalculable value.

The remedy for this deplorable state of affairs must, necessarily be left to the wisdom of Congress. But the navy, with no other interest in the question save that dictated by the highest sense of patriotism, discharges an imperative duty, in urging as a military necessity, the rehabilitation of our mercantile marine.[35]

Just as Admiral Luce advised, it was the Congress, not the Navy, that placed the American mercantile marine on healthy enough footing to fight the Pacific War, securing the strength of the commercial and political points of the logic trident when it was necessary to call on American sea power. While the Navy identified the necessary characteristics of the merchant vessels it needed as auxiliaries (for instance identifying the need for fast fleet oilers during the fleet problems), it had little hand in determining the policies that would be used to build the United States merchant fleet.

What the Navy did know that it needed from the merchant marine— the commerce portion on Holmes's and Yoshihara's sea power logic trident— stemmed from its understanding of the requirements of a Pacific War to defend the Philippines from Japan. The requirements of such a campaign, especially the problems of distance and logistics, would by necessity have to be solved through the merchant marine. The logistics problems, as planners saw them, were:

> To project its power to the western Pacific the United States would have to over-come unprecedented difficulties. The Blue navy was habitually stationed in the Chesapeake. To reach the main war zone, it would have to travel 14,000 miles via the Atlantic and Indian oceans or 19,700 miles by way of the Straits of Magellan, the latter distance equivalent to nearly 80 percent of the earth's circumference. After the Panama Canal opened in 1914, the cruise shrank to a still formidable 12,000 miles. Even the naval bases proposed for construction in California and Hawaii would lie 7,000 and 5,000 miles from the Philippines. Simply getting to the battlefront would be a heroic effort logistically. The destruction of a weary Russian fleet in 1905 after an arduous voyage from the Baltic to the Battle of Tsushima was a somber omen.
>
> A fleet's power was estimated to erode by 10 percent for each thousand miles it cruised from its base. Wear and tear, prolonged absences for repairs, and tropical growth that fouled hulls and reduced speed by several knots within a few months of leaving dry dock would inevitably erode American naval superiority. The fighting strength of the fleet would decline steeply in conformance with the "N-squared rule," which decreed that relative power in battle was proportional to the square of

the number of ships (or guns) in action. These equations of distance and power underlay the famous five-to-three battleship tonnage ratio negotiated by the United States and Japan in the Washington Treaty of 1922 to equalize their combat power in the western Pacific. (Extension of ship ranges by conversion to oil and later by refueling at sea eased the penalty of distance, but the problem worsened again in the air age because planes had action radii of hundreds, not thousands, of miles). Furthermore, the long haul from America's arsenals and coal mined dictated employment of five to ten times as many merchant vessels as Japan to maintain equal strength in the war zone, and they would be exposed to attack during much of their course whereas Japan could service its outposts free of risk until U.S. warships reached the theater.[36]

Thus there were two main concerns: mitigating the wearing of the fleet from long distance travel and requiring almost an order of magnitude larger merchant marine than Japan just to support equal strength in the war zone. The latter problem would be substantially assisted by modern auxiliaries, but it was really a problem of quantity, not quality, and certainly did not require revolutions in ship design. However, the first problem required a much more technical solution. The U.S. Navy, unlike the other belligerents in the Pacific War, conquered the challenge of distance and the 1,000 nm/10 percent rule:

> Americans excelled at maritime logistics. The hundreds of Liberty ships that were launched to haul men and goods were protected, as prescribed in Plan Orange, by convoying, ports of refuge, and anti-raider patrols. (Japan failed effectively to do the same, to its regret.) The United States, alone among naval powers, mastered the art of carrying bases with the fleet. Logisticians had been compelled by vetoes of overseas dockyards and the dilemma of Pacific distances to invent giant colliers in 1908–14 and the Western Base project in the 1920s. They designed modular advanced base units for assembly by special construction battalions (Seabees). Halsey called the bulldozer one of the three decisive weapons of the war. The floating service depot was perfected. The war lasted long enough for delivery of the huge floating dry docks that had worried two generations of planners, although new damage-control techniques and antifouling paints rendered them somewhat less critical. A U.S. Navy unaccompanied by mobile bases could probably not have defeated the Japanese at sea once they adopted a defensive posture. In contrast, the Royal Navy was helpless to return after the loss of Singapore until the Americans provided logistical support in 1945.[37]

The development of the floating dry dock, serving as the floating base that the fleet needed, will be discussed below. However, the passage above conveys the extreme importance with which the Navy took logistical matters. Not only did they need to field advanced weapons, they needed to field advanced logistics systems. The Navy could build floating bases and deploy naval engineering battalions, but they could not take the lead in developing every logistical requirement the Pacific War needed. Just as Admiral Luce had predicted, it was the legislative branch that would take up the essential commercial shipping piece of sea power. Fortunately for the United States, Congress took major action to develop the merchant marine just before the beginning of the war, and just in time. Historian John Butler explains:

Again Congress studied the lagging state of commerce and defense. This time it came up with the most enduring legislation of the century. On June 29, Roosevelt signed the Merchant Marine Act of 1936, abolishing the Shipping Board Bureau and creating an independent agency, the U.S. Maritime Commission (MarComm). MarComm was authorized to regulate American ocean commerce, supervise freight and terminal facilities, and administer government funds for the construction and operation of commercial vessels. Any American company that was three-quarters owned by American citizens could apply for financial assistance in having new ships built in American yards. If a ship was put into what MarComm defined as a service essential to foreign trade, a construction differential subsidy" would be paid, based on the difference between the potential costs of foreign and home construction. Operating companies were offered twenty-year loans amounting to 75 percent of a vessel's purchase price. Aboard ship, tighter licensing requirements were specified, the three-watch system established for all departments, and specifications set for more comfortable crew quarters. Subsidized freighters were required to have fully American crews, passenger ships 90 percent American. (Filipino and European stewards held greater appeal for paying passengers.) To modernize and keep the merchant marine efficient for national defense, ships constructed and operated under the American flag were ready for conversion to naval auxiliaries. Further, the 1936 act provided funds to support the training of officers for manning such vessels.

The shipyard labor force was down to twenty thousand workers. There were no more than ten yards that could build ships longer than four hundred feet, and half of the forty-six slipways were tied up in naval contracts. During the fifteen years from 1922 to 1937, only two dry cargo ships, a few tankers, and twenty-nine subsidized passenger ships were built. Roosevelt appointed Joseph P. Kennedy as first head of the Maritime Commission in 1937. Kennedy was quick to recognize the end of a failed era, stating that in twenty years the U.S. government had spent $3.8 billion on *not* getting a healthy merchant marine. He proposed a ten-year program to construct five hundred ships, and soon after, his eyes on loftier political assignments, he departed from the agency. His replacement was a commission member, Rear Admiral Emory S. Land, who moved to get congressional authority to build 150 vessels as soon as contracts could be written.

In March 1938 MarComm established the Cadet Corps and assumed the responsibility of selecting and training men through assignments aboard subsidized merchant ships. Richard R. McNulty, a captain (later rear admiral) in the Naval Reserve, was named as first supervisor of the corps, a post he held for twelve years. A graduate of the Massachusetts Nautical School, McNulty was charged with establishing a competitive selection process for formal schools to be located on the three coasts. Within a year, and with little more base of operations than was afforded by the ships they traveled on, more than two hundred cadet-trainees were working toward licensed positions on federally funded merchant ships.[38]

With MarComm, subsidized ships and crews, and a new Cadet Corps training pipeline, the institutional architecture for a flowering American merchant marine was in place. However, even with Admirals Land and McNulty as major MarComm players, MarComm was certainly not a U.S. Navy creation. It was overwhelmingly a civilian-initiated organization. So why were civilians taking such a navalist position? Gibson and Donovan argue that the navalist understanding of sea power was not limited to those in uniform:

> In its first year the commission also began a study of the industrial economics of the maritime industry.... The report that summarized the study's findings, *An Economic Survey of the American Merchant Marine*, was published in November 1937....
>
> While *An Economic Survey* concentrates on economic analysis, it is no mere academic exercise.... The introduction opens with a remarkably direct and well-informed description of the business of shipping and a warning:
>
>> Shipping is our oldest industry; it is also one of our most complex. Shipping, in the first place, is not a business in the usual sense of the word. It is, so far as the United States is concerned, an instrument of national policy, maintained at large cost to service the needs of commerce and defense. It is operated, meanwhile, as a private enterprise. We like to think of shipping as an example of individual initiative, sustained by investments and capable of being operated at a profit. In practice, however, the industry requires substantial Government support to survive. That entails some measure of Government control, which in turn means inflexibility, curtail of investment, and perhaps in the end an increase in the needs for subsidies.
>
> The introduction's account of the reasons for providing government support for the shipping industry is equally concise: "Careful examination of the arguments advanced on behalf of an American Merchant Marine indicates that there are only two sound considerations that justify the expenditure of public funds to maintain a foreign-going fleet by the United States: One of these considerations is the importance of shipping as a factor in the preservation and development of foreign commerce; the other is the relationship that exists between merchant vessels and national defense. Upon these two considerations must rest the case for the maintenance of a subsidized shipping establishment in the international carrying trades."[39]

In this study, MarComm identified and committed to the navalist belief that commerce and war fighting on the high seas were intimately linked, and that a robust Navy requires a robust merchant marine. Moreover, MarComm stated that subsidies to merchant vessels were not only necessary but practical measures to develop the robust merchant marine required by the United States, and the merchant service needed by War Plan Orange.

With this understanding of the importance of the merchant marine to national power and justification for government regulation and subsidies.

The United States had turned a corner, embarking on a new road that would carry the maritime industry to the end of the twentieth century. The Merchant Marine Act of 1936 remained true to Woodrow Wilson's vision of a powerful national industry, but it downplayed Wilson's economic rationale. It adopted almost word for word the statement of purpose in the Merchant Marine Act of 1920, substituting only that the goal was a merchant marine to carry "a substantial portion" of the nation's shipping needs instead of "the greater portion." To achieve that end, it created different subsidies to support both construction and operation of oceanic liners. The act did not support ships on coastal or inland routes, for these faced no foreign competition. Nor did it apply to tramp steamers or bulk carriers, which managed to compete internationally with varying degrees of success. It was liner service, that was the most competitive, the most volatile, potentially the most lucrative, and in time of war the most useful to American military needs. The government therefore agreed to pay shipbuilders a construction subsidy equal to the difference between their costs and those of foreign firms building ships abroad, up to 50 percent of the price of a vessel. And it agreed to pay shipping firms for the difference between their operating costs and those of firms operating ships under foreign registry, up to 75 percent of yearly totals.[40]

These political and economic actions proved perhaps as devastating to the Axis powers in World War II as the American military, for they were the forces that woke the sleeping giant from its slumber. By energizing the economic and political tines of the sea power trident, the United States dealt its enemies an unexpected and critical blow:

> It is hardly surprising that the Germans and Japanese did not anticipate the flood of manufactured goods that poured forth from America's factories and shipyards to support its war effort. A country that had constructed only 1 million tons of merchant shipping in 1941 built more than 17 tons by 1943. By the war's end in 1945, a workforce of 4 million men and women had built 5,000 ships at a cost of some $12 billion. The speed with which this was done and the scale of the effort still defy comprehension. When operating at their peak rate of production, America's shipyards were capable of reproducing the entire world's prewar commercial tonnage in less than three years.[41]

The sheer magnitude of the merchant marine buildup for World War II is one of the most amazing industrial feats in military history, far eclipsing the buildup of warships during that time. The sheer industrial might of the United States makes buildups like these look easy. Seemingly as soon as the nation needed them, industry built thousands of ships essentially from scratch. This is not true. This massive buildup required many things before the war started in order to succeed: the funding, the organizational structure of Mar-Comm, and the navalist belief that the United States would need a truly world-class merchant marine in order to fight Japan.

Also needed were businesses able and willing to operate the merchant fleet. Even though the merchant marine was necessary to the defense of the nation, they were not manned or operated by naval personnel. Along with the rapid American maritime expansion came an expansion of private involvement in maritime commerce:

> With approximately forty-five hundred commercial vessels—roughly 60 percent of the world's commercial tonnage—America's merchant marine was then [at the end of World War II] the largest in the world. These ships were operated by about 130 private companies with extensive experience in maritime operations, considerable financial resources, and a determination to say in business. This expansion amounted to a threefold increase in the number of operators and a fourfold increase in the number of ships since 1939.[42]

Though the navalists were often government employees or officials, and the Navy and MarComm federal organizations, sea power was never solely the province of government. The businesses, ship owners, and merchant crews were instrumental in developing and deploying American sea power for the Pacific War.

Even with the monumental magnitude of the effort to build the American merchant marine just before World War II, the shipping necessary to win the wars in the Pacific as well as the Atlantic couldn't be built entirely from scratch. The Merchant Marine Act of 1936 was extremely important, but it by itself was not enough to win. The remaining merchant marine hulls from World War I were equally necessary. Winning the Pacific War required, not only the navalists' call for merchant ships just before the war, but required as much effort as could be mustered over the prior 60 years:

> The traditional narrative of World War II shipping features the Liberty ships, Victory ships, and tankers that gave the war's crash shipbuilding program its public face. These emergency vessels, so the story goes, issued from American shipyards in sufficient numbers to replenish the losses to German submarines in the North Atlantic. There is more than a little truth to this picture. But most of America's shipping losses were not Liberties and Victories but stalwart veterans of the World War I shipbuilding program. Almost half of the American merchant vessels lost in World War II—306 ships—began life in the World War I Emergency Fleet. Another 11 percent either predated that conflict or came into the American fleet from other sources. Only 32 percent of the merchant ships lost in World War II were products of the war's shipbuilding program.
>
> One reason was timing. The United States began World War II with an old fleet, and 1942 was by far the most dangerous year of the war for merchant shipping. Half the American ships lost or damaged during the war succumbed before the end of 1942. The U.S. Maritime Commission building program delivered 760 ships in 1942, and fewer than a thousand during the total prewar buildup beginning in 1937. In comparison, it produced almost five thousand vessels in the last three years of the war.[43]

World War II required the U.S. Navy to fight massive campaigns world-wide and its merchant marine to serve on every sea. However, the Pacific War was unique in that it required not only the ships (both warships and merchantmen), but substantially all of the heavy maritime industry to be developed simultaneously with which to prosecute it. Where the Atlantic and Gulf regions already had well-developed ports and shipyards with which to supply the American war effort, the Pacific coast was far less equipped:

> The West Coast merchant marine supported the Pacific campaign from Pearl Harbor to Tokyo. It got its seamen by the same mechanisms feeding East Coast shipping. But its ships and ports were a different story. From 1940 through 1945, shipping from American ports on the North Atlantic increased by 240 percent, while that from South Atlantic and Gulf ports increased by 180 percent. But export freight from West Coast ports increased in the same period 1,487 percent, surpassing by almost 100 percent the volume of traffic out of the South Atlantic and Gulf ports and approaching that of the North Atlantic ports.
>
> The West Coast accounted for 7 percent of U.S. export freight shipping in 1940, 34.5 percent in 1945. World War II drove the creation of the Pacific merchant marine the way World War I had driven the Wilson/Hurley vision of an American bid for the North Atlantic market. This achievement taxed the entire U.S. transportation infrastructure, from inland and coastal shipping to railroads and rubber-wheeled vehicles (cars, trucks, buses). The nexus of overseas shipping, the ports, required a transformation of existing infrastructure and the creation of entirely new facilities where none had existed before.
>
> Shipbuilding on the West Coast experienced similarly explosive growth. In 1941 the West Coast was producing less than a third of the ships launched in the United States, hardly more than half as many as the East Coast. The following year, it was producing more than any other region. The year after that it built more than half of the national total. By war's end, the West Coast had built 47 percent of the nation's new tonnage, more than that constructed in Wilmington and the other East Coast shipyards.[44]

Of course, the common training pipeline for merchant seamen for both the West and East Coast merchant marine required as much development and the physical shipbuilding program. At the beginning of the Pacific War, the trained seamen capable of operating a world-class and world-spanning merchant marine was far too small to prosecute war effectively. Knowing full well that ships need to be manned as well as constructed, the merchant officer and deck hand training program was rapidly expanded to match the ambitions building program.

> Only about fifty-five thousand merchant seamen and officers were sailing in December 1941. Hundreds of thousands would be required to crew the ships that would carry the war to the enemy. By the spring of 1942, an average of forty-five ship sailings a month suffered delays from lack of crews. The War Shipping Administration

(WSA) sought out and recruited experienced seamen engaged in other walks of life and set up training programs for new ones. The United States Merchant Marine Cadet Corps, authorized by the Merchant Marine Act of 1936 and created in 1938, took up residence at the U.S. Merchant Marine Academy at Kings Point, New York, in January 1942. Other institutions around the country joined, finally producing, between 1938 and December 1945, 31,986 officers, 7,727 radio operators, 150,734 unlicensed seamen, 5,034 junior assistant purser–hospital corpsmen, 2,588 junior marine officers of the Army Transport Service, and 64,298 other graduates of specialized training programs who either learned anew or upgraded their maritime skills. In all, 262,474 graduates of these various programs qualified themselves to operate the country's merchant vessels.[45]

Just as the Navy didn't get everything right in planning for the war aspects of the Pacific War, neither were they always correct in what they relied upon from the merchant service:

The [War Plan Orange] planners hoped to convert fast passenger ships into supplemental fleet carriers (XCVs) at the rate of one per month. But U.S. liners were unimpressive in speed and range, and it was found that elevators and other gear could not be easily installed in them. Besides, they would be needed as troopships. The XCVs soon faded from Orange Plans. The abundant escort carriers (CVEs) of World War II performed supporting missions unlike those hypothesized for the fleet XCVs of the 1920s.[46]

Here is an example of a requirement given originally to the commercial arm of the trident of sea power ultimately giving way to the military arm due to experimental evidence and assessment. Therefore, developing the merchant and military tines were not accomplished in separate vacuums. All tines of the marine trident were required to be integrated because fighting the Pacific War was the responsibility of sea power in its entirety, not simply its military expression.

With the ports, ships, and men, the commercial service was ready to serve alongside its warship cousins to fight and win the Pacific War. With the innovative new designs and operations techniques, both the fighting and merchant navies were ready to answer America's Pacific call. In preparing for the Pacific War, both civilian and uniformed navalists brought the three tridents of sea power to defeat the Japanese Empire in the war the navalists had predicted over a half century earlier.

The Washington Naval Treaty and the Base Problem: A Case Study

The triumph of the navalists was not easy and not without serious setbacks. Perhaps the most dangerous problem the navalists encountered was the

political decision to sign the Washington Naval Treaty and the extremely damaging problems it caused to military sea power readiness. This incident is especially useful for our study of space power because it shows how political
constraints can drive innovation and change in order to keep the power logic
correctly balanced. The subject of this case study, to examine how Mahan's
ideas can be used to inform operational concerns at the grammar of war level,
is the Washington Naval Treaty and the construction of the "Treaty Navy" in
the 1920s and '30s. Even though this study uses Mahan's ideas in its classic sea
power context, we will conclude with some discussion on the space treaties
that are currently in effect to see that these principles will remain constant in
the sea or space realms.

Commander John T. Kuehn's *Agents of Innovation* describes how the
navy confronted the political decisions made by the signing of the Washington
Naval Treaty that caused it to change an important assumption of its Pacific
war plans. Kuehn explains that after World War I, "U.S. naval strategy revolved
around the successful application of sea power, which was universally understood by navy officers to be comprised of a fleet, domestic and overseas bases,
and a robust merchant marine."[47] These components correspond to the elements of the grammar of war, since the merchant marine is a proxy for the
maritime trade of the nation, which generates the treasure needed to prosecute
military operations. The navy also realized that a deficiency in one element
would necessarily demand to be compensated by increasing a different element
in order to maintain superiority over an adversary.[48]

The deficiency mandated by the Washington Naval Treaty of 1922 was
from Article XIX, known as the Fortification Clause. In it, the article specified:

> The United States, the British Empire, and Japan agree that the status quo at the
> time of the signing of the present Treaty, with regard to fortifications and naval
> bases, shall be maintained.... The Maintenance of the status quo under the foregoing
> provisions *implies that no new fortifications or naval bases shall be established....*[49]

Article XIX struck at the very heart of the U.S. Navy's Pacific strategy. By not
allowing further development of America's Pacific bases, the treaty severely
hampered the navy's original plans to retain strategic access in the Pacific—
specifically the defense of the Philippines and the "Open Door" trade policy
with China.[50] Admiral Bradley Fiske, writing in 1916, detailed the navy's base
strategy:

> The Pacific Ocean is so vast, and the interests of the United States there will some
> day be so great, that the question of establishing naval bases, in addition to bases
> at Pearl Harbor, the Philippines, and Guam, will soon demand attention.... A mod-

erately far-seeing policy regarding the Pacific, and a moderately far-seeing strategy for carrying out the policy, would dictate the establishment and adequate protection of bases in both the southern and the northern regions; so that our fleet could operate without undue handicap over the long distances required. The same principles that govern the selection of positions and the establishment of bases in the Atlantic apply in the Pacific; the same requirements that a base shall be near where the fleet will conduct its operations—no matter whether those operations be offensive or defensive, no matter whether they concern direct attack or a threat against communications.[51]

As previously stated, naval planners at the time generally assumed that a fighting fleet would lose 10 percent of its combat effectiveness for every 1,000 nautical miles it operated away from its home base.[52] To avoid this degradation, the navy had intended, before the treaty, to build a major naval base on Guam as well as reinforce the established base on Luzon. An early War Plan Orange (the war plan against Japan in the Pacific) called for the U.S. Navy to defeat the Japanese fleet in a decisive battle off of Guam and then conduct relief operations toward the Philippines if necessary.[53] Since fleet logisticians believed that a fleet could not operate effectively beyond 2,000 miles from a major naval base, many bases were needed besides Hawaii and the Philippines to give the fleet full range of movement in the Pacific.[54] These bases were not developed to the necessary potential before the treaty was signed, and no further development was allowed, making these anticipated bases useless. Indeed, many navy leaders learned of the Fortification Clause only after ratification, and thus began a crisis of naval operational strategy.[55]

Without these bases, the navy's "pre–Treaty" ships would not have sufficient strategic access to the Pacific to effectively project power and carry out War Plan Orange. One of the necessary elements of the grammar of war— bases—was kept below its necessary strength by the treaty. The navy was faced with an operational problem. How could the navy effectively create and maintain strategic access to the Pacific Ocean for a fleet able to defeat the Japanese navy far away from the support of any major naval bases? The navy would have to use the grammar of war to rebalance the military tine of the logic of sea power, reexamine how strategic access was to be generated, and take up the slack from the loss of bases by boosting war's other elements—and finally by redefining the concept of the base itself.

Captain Dudley Knox was among the first naval officers to note the importance of the Fortification Clause to the navy's Pacific strategy. Writing in 1922, Knox said that the "sacrifices [the U.S.] has made respecting Western Pacific bases" were of "greater importance" than the much better known limits on capital ship tonnage and allowable ship types the treaty represented in his passionately named book *The Eclipse of American Sea Power*.[56] To compensate, Knox argued that the U.S. would have improve its military shipping element, her battle fleet, by building more fighting and auxiliary ships. However, this strategic option was of limited value for two reasons. First, any attempts to increase the cruising radius of fleet and auxiliary units would result in design tradeoffs that would limit the new ships' ability to fight effectively against similar classes of ships optimized for firepower rather than endurance. Second, any shipbuilding program initiated by the U.S. would be matched or exceeded by an unrestricted Japanese building program, which could economize their fleet strategy with their current base structure. These two reasons made it difficult, if not impossible, to rebalance the grammar of war by substituting fleet units for bases.

Increasing the treasure available to the navy by boosting maritime trade through merchant marine expansion would prove equally difficult. Without a competitive naval presence in East Asia, any increase in maritime trade with China and other nations would be at the mercy of the Imperial Japanese Navy. In the event of war, American maritime trade in the Pacific would be jeopardized and, in all likelihood, ended immediately—drying up any treasure to the navy a larger merchant marine could provide. Strategic access would be cut, and American sea power would decline in logic as well as grammar. Also, the well-known predisposition of democracies to spend only sparingly on military and defense matters would likely cause any increase in revenue through maritime trade to be spent on anything other than defense of strategic access to the Pacific. It would appear that attempting to substitute treasure for bases wouldn't solve the Navy's operational dilemma either.

Captain Frank Schofield underlined the severity of the treaty's military operational sea power problem in 1923:

> Sea power is not made of ships, or of ships and men, but of ships and men and bases far and wide. Ships without outlying bases are almost helpless—will be helpless unless they conquer bases and yet the treaty took from us every possibility of an outlying base in the Pacific except one [Hawaii]; we gave our new capital ships and our right to build bases for a better international feeling—but no one gave us anything. *Manifestly the provisions of the treaty presented a naval problem of the first magnitude that demanded immediate solution.* A new policy had to be formulated which would make the best possible use of the new conditions.[57]

The grammar of politics stabbed the grammar of war in the back with the treaty, at least to the Navy. But the problem remained. What policy could be formulated that would make the best possible use of the treaty's conditions? Expanding the fleet and investing in commerce were not, by themselves, certain to bridge the gap in military sea power in the Pacific from the gutting of the base element's potential. The answer would come from a reexamination of the concept of the base itself.

Navy civil engineer A.C. Cunningham stumbled onto the solution in 1904 with his concept of the mobile base.[58] Cunningham's "moveable base" would be "composed of sectionalized floating dry docks, colliers, ammunition, repair, supply, and hospital ships [that] would move with or behind the fleet, and would offer all the essential services required of a base."[59] With the mobile base, the Navy could retain all the services required of naval bases without requiring them to be built on land and, most importantly, without violating the Fortification Clause of the Washington Naval Treaty. When confronted with the treaty, planners on the Chief of Naval Operations staff found Cunningham's idea and pressed it into service to rebalance the grammar of war in the Pacific by giving the base element much of its status lost from the treaty. The evolution of the mobile base concept eventually produced the Fleet Base Force, one of the four component fleets of the U.S., which included repair ships, store ships, hospital ships, refrigerator ships, and (by the end of World War II) six Advanced Base Sectional Docks—movable dry docks rated to hold and repair ships up to 90,000 tons.[60] The Fleet Base Force was truly a mobile base brought to life. When confronted with the loss of the potential of fixed bases to ensure the U.S. Navy's strategic access to the Pacific, the Navy managed to attach the bases to the fleet and saved the fighting potential of sea power through a new understanding of the elements of the grammar of war.

The Navy ultimately reacted to the new conditions under the treaty in a three-point plan of battleship modernization (to increase mobility and range), robust building programs to the limits of the treaty and for ship classes not regulated under the treaty (including maximizing aircraft carriers, destroyers, and submarines), and the development of the Fleet Base Force.[61] These three advances in the 1920s and '30s were the genesis of the blue water fleet that held off and finally devastated the Imperial Japanese Navy in the Pacific War. These advances, especially the Fleet Base Force, were spectacular innovations never before seen in naval warfare. However, they did not invalidate the grammar of war of sea power—they were simply new expressions of the tried and true concepts. Interwar U.S. naval innovation was truly a revolution within the form. Commander Kuehn says it best:

> The Navy did not make a radical change to its paradigm of sea power during the interwar period. Instead it tried to adjust its solutions and designs to fit an existing paradigm or conception of sea power that was rooted in the teachings of A.T. Mahan. In doing this the Navy changed the boundaries, but not the essentials, of the paradigm. Fleets, maritime commerce, and basing were all still regarded as essential to the sea power equation. However, the definition of just what "basing" really

meant had been expanded to include mobile bases. Both fixed and mobile bases were needed in the strategic environment of the Pacific. Although the paradigm did not change, the fleet did. This change was accelerated by the anomaly caused by the fortification clause to the Navy's traditional conception of sea power. It became axiomatic that in wartime mobility, not just of tactical and operational units but for strategic logistical capabilities, was vital. The fortification clause caused the U.S. Navy to learn how to become a global navy.[62]

This case study shows how the Mahanian model doesn't break down when confronted with new input at either the logic or grammar, strategic or operational, levels. The Washington Naval Treaty was a classic sea power heresy that neglected the logic of sea power—maritime wealth through strategic access, by elevating the grammar of politics (good international feeling) by sacrificing the grammar of war (bases needed to ensure strategic access). However, as the case study also showed, the Mahanian sea power model was able to inform planners and strategist on how the Navy could respond successfully to meet the needs of the new Treaty-inspired strategic environment. Through innovation inspired by Mahanian sea power theory, the United States in 20 years went from standing on a politically driven precipice that threatened the eclipse of American sea power to fielding a global blue-water naval fleet with strength unparalleled in the history of warfare. A robust model, indeed.

But why should space planners and strategists care? These events occurred over 60 years ago, on the ocean. Even if Mahanian sea power theory can be successfully employed in a space power setting, how can reviewing the Navy's operational reaction to the Washington Naval Treaty help space planners? Consider this. Article II of the treaty on *Principles Governing the Activities of States in the Exploration and Use of Outer Space, Including the Moon and Other Celestial Bodies* (Outer Space Treaty), signed by the United States in 1967, states:

> Outer space, including the Moon and other celestial bodies, is not subject to national appropriation by claims of sovereignty, by means of use or occupation, or by any other means.[63]

Article IV of the same treaty continues:

> The establishment of military bases, installations and fortifications, the testing of any types of weapons and the conduct of military maneuvers on celestial bodies shall be forbidden.[64]

Uncanny similarities to the Fortification Clause of the Washington Naval Treaty cannot go unnoticed.

Established international treaties go beyond just impinging on the freedom to exercise the grammar of war in space power. The *Agreement Governing the Activities of States on the Moon and Other Celestial Bodies* (Moon Treaty of 1979), not signed by the United States, attacks the grammar of commerce as well. Article 11, Paragraph 3, makes state-owned or private property on celestial bodies illegal by stating:

> Neither the surface nor the subsurface of the Moon, nor any part thereof or natural resources in place, shall become property of any State, international intergovernmental or non-governmental organization, national organization or non-governmental entity or of any natural person.[65]

Paragraph 5 of the same article mandates space natural resource extraction the exclusive purview of international governments:

> State Parties to this Agreement hereby undertake to establish an international regime, including appropriate procedures, to govern the exploitation of the natural resources of the Moon.[66]

It is clear from even a cursory review of space treaties that they could interfere with the development of space power under the Mahanian model. The above selections of the Outer Space Treaty and Moon Treaty prove that these treaties can be considered, under the Mahanian model, to be examples of the political heresy, raising the grammar of politics over that of commerce and war together, and completely ignoring the Logic of Space Power altogether. Much like the Washington Naval Treaty, diplomats may have sacrificed much needed freedom of activity in space for nothing more than good international feeling. Whether the General Theory of Space Power advises us to repudiate these treaties as heresies or force military planners to adjust the grammar of war and business executives to adjust the grammar of commerce to accommodate the politicians, only time and study will tell. What should be extraordinarily clear is that the experiences of the naval planners of the 1920s and '30s are extremely relevant to today's space power strategists. Using the Mahanian space power model described here, space strategists may be able to advance their art considerably.

Preparing for the Navalists' War through the General Theory of Space Power

Through the roughly sixty years the navalists had to prepare the Navy for the Pacific War, breakthroughs occurred in every facet of naval warfare. New machines, new tactics, and new organizational strategies that had great impacts throughout both the Grammar and Logic of Sea Power transformed the Navy from a Civil War relic into a global powerhouse. Now that we have explored how the navalists accomplished this transformation, let us explore their actions through the General Theory of Space Power. Recall the organizational revolution enacted by the navalists:

> On balance, organizational factors played a positive role in influencing innovation during the [interwar] period in the U.S. Navy. The factors of most importance were the interrelationships among Navy organizations, how members were assigned to these organizations, and the internal structures of the organizations themselves. The General Board played a central role as an organizational entity that facilitated innovation because it was here that the implementation of [Washington Naval Treaty] policy intersected with force and ship design. This intersection was further positively influenced by the composition and structure of the Board. Members brought their unique experiences and relationships to the Board and continued to refer back to the Board as they went to other billets. Often their next job or their previous job involved close association with the General Board, for example Admiral Pratt's assignment as president of the Naval War College after his stint on the General Board in the 1920s or Captain Schofield's similar assignment from the Board to Chief of Naval Operations' chief of war plans. This "job-shuffling" favored, rather than discouraged, organizational collaboration in the interwar Navy. Also, most of the officers assigned to the General Board had already "bought into" the Navy's experimental approach to problem-solving that was taught and best exemplified by the War College curricula and practiced during Fleet Battle Problems.[67]

Recall that development can occur on two different levels, the grammar and logic of an environmental power. Grammar developments concern building bigger, better, and more building blocks of power—tangible tools and vehicles that can travel and operate in a particular realm. Logic development, alternatively, deals with finding new and better ways to use those tools to support a goal. Economist Joseph Schumpeter identified five different paths of economic development, which can also be used to describe environmental power development. The first, introduction of a new element (a production, shipping, or colony element) essentially is development through the introduction of a new piece of hardware. The second path, introduction of a new method of production or handling of an element, essentially deals with using existing elements in new ways. Path 3 development is the opening of a new market, defined in the General Theory of Space Power as combining elements into a new level of access. Path 4 development is a straightforward conquest of a new source of supply, access to a new cache of strategic materials that changes the zero-sum status quo. The last path to development, Path 5, is the reorganization of an industry or organization that stimulates new logic and grammar ideas with which to further fuel power development.

New elements are combined in new ways to form new accesses and dominion over the environment, the ultimate expression of the grammar of power. Logic takes these new accesses and uses them through transformers to develop new abilities to use applied power towards an objective, the ultimate expression of the logic of power. In grammar, elements are often the most important ingredient, with combinations being new recipes. In logic, the transformers are paramount. Military transformers are tactics, techniques, and doctrine, which direct the use of military equipment for military purposes. Commercial transformers tend to be business plans, which use civilian equipment for profit, producing wealth. Political transformers tend to be treaties, agreements, and goodwill gestures, which generate prestige and soft power. In this framework, the navalist programs can be better understood and adapted for use in generating space power.

THE NAVAL WAR COLLEGE–
GENERAL BOARD–OPNAV TRIANGLE

Perhaps the most important development of the navalist era was reorganizing the Navy for development. Creating the Naval War College to stimulate idea generation, the General Board to consider these ideas, and the Office of the Chief of Naval Operations to test and deploy the fruits of these ideas was a master stroke of organizational revolution in innovation.

Under the General Theory of Space Power, the NWC–GB–OpNav structure is a prime example of Path 5 development. The simple retooling of the naval leadership structure was not based on any new technology or grammar element at all. The new structure simply allowed greater generation and development of ideas at the logic level. The triangle did not develop the submarine, airplane, aircraft carrier, or floating dry dock, but it most certainly led to the development of advanced ideas on how to use these new potential instruments of sea power. Begun by studies at the Naval War College and associated war games, ideas went to the General Board for consideration and review and, ultimately, to the fleet (through OpNav) for live testing (with feedback to the NWC and GB) and deployment. The triangle led directly to a renaissance of new military transformers at the logic level of sea power and provided the structure with which to further develop and refine these ideas—as well as the elements needed to employ them—into extremely advanced new sea power abilities. The navalist reorganization of the Navy stands as perhaps the most important innovation of the navalist era and a museum quality example of Path 5 development along the military power point of the logic delta. However, in order to truly release the innovative power of the triangle organization, it needed sufficient grist to evaluate and ponder. Thus we have the next great engine of development, the Fleet Problems.

THE FLEET PROBLEMS

The interwar Fleet Problems were the limbs to the NWC–GB–OpNav triangle's brain in Navy development. It was in the Fleet Problems that the fleet was introduced to the innovational military transformers at the logic level and confronted the cold, hard reality of sea operations at the grammar level. The Fleet Problems tested both transformers at the logic level and elements and their combinations at the grammar level. Therefore, the Fleet Problems assisted the development of both sea power logic and grammar directly. The Fleet Problems stimulated development through Path 2 development as the fleet struggled with refining and incorporating new ways to handle new elements under military conditions. The problems also affected grammar through Path 2 development, uncovering new combinations of elements to generate new accesses. The lessons learned from the Fleet Problems also drove warship and aircraft design, stimulating Path 1 development as new elements requested by the fleet came online.

First, the Fleet Problems tested the military transformer ideas developed by the triangle under sea conditions. Examples include the first use of naval aviation as scouts to the prewar striking arm of the Fast Carrier Task Force as well as the procedures necessary for safe and effective refueling replenishment at sea. While some ideas looked good on paper in the classrooms of the Naval War College or the foyer of the General Board, only their relevance on the decks of the fleet afloat really mattered to development. The Fleet Problems provided an all-important laboratory with which to test new transformer innovation at the logic level.

The next essential role the Fleet Problems played was at the grammar level. The Fleet Problems dealt directly with military equipment and their uses, at the element and combinations level of the Grammar Delta. First, the combinations of elements were tested to see if naval systems worked as they were intended through Path 1 and Path 2 development. Some advances, such as the airplane and aircraft carriers, worked wonderfully and generated successful naval advances. Others, such as the Navy's dirigible program, were deemed failures and abandoned. Without the hard and realistic experience the Navy earned from the Fleet Problems, advances from neither the Logic nor Grammar Deltas of sea power would have been ready for use in the Pacific War.

Here we must also state that the dedication of the Navy to development rather than readiness (as exhibited in the Fleet Problems) was a superior strategic decision, and one that any student of space power must also ponder as we enter a period of rapid technological and industrial change. The Navy consciously sacrificed readiness by engaging in elaborate and expensive Fleet Problems, and continued to do so consistently until it was clear that war was coming. Due to the Navy's superior strategic foresight, the Pacific War Navy was ready when the nation called upon it.

Technology Procurement

Of course, environmental power cannot be exercised without the proper equipment. Therefore, while logic development is extremely important, it by itself is impotent without a strongly developed grammar power base. The Navy's organizational triangle and the Fleet Problem systems were useless without new ships and planes, and these too had to be developed. At the beginning of the navalist era, there was no such thing as an airplane nor aircraft carrier, fast troopships were unheard of and floating dry docks as fanciful as invaders from Mars. Navalists had to focus on the Grammar of Sea Power as well as its logic.

The development of the airplane innovation into naval aviation encompassed many path developments. Organizationally, Path 5 development began with the activation of the Bureau of Aeronautics. Path 1 development began when the Navy purchased its first Wright aircraft. Path 2 development commenced in tandem with advances in automotive engine technology, avionics, and aircraft carrier technology.

It is important to note the Navy stimulated both Paths 1 and 2 development through its acquisition policy. Throughout the interwar years, the Navy (intending to incorporate new lessons learned from the Fleet Problems as well as capitalizing on rapid advances in relevant technologies and industries) practiced ordering military aircraft in small batches. Between World War I and World War II (excluding contracts that extended into World War II), the Navy acquired 1,917 combat aircraft of 27 types, for an average production run of only 71 aircraft per type.[68] "Small batch acquisition" allowed the Navy to incorporate new designs quickly and ensured that innovation (from both Grammar and Logic Deltas) was deployed to the fleet as quickly as possible. On balance, it appears that small batch acquisition is a key necessity for a force dedicated to development as opposed to readiness. As the aircraft acquisition contracts show, when the Navy fought World War II, the amount of new aircraft designs decreased but the numbers of aircraft produced increased by orders of magnitude. Most of even the advanced aircraft of Pacific War were already in advanced design stages by its beginning. In order to stimulate innovation, small batch acquisition is key to keep Path 1 and Path 2 development continuously ongoing.

The aircraft carrier's structural development matured in many ways like the aircraft. Initial Navy carriers were converted merchant ships (USS *Langley* CV-1 was originally the collier *Jupiter*), then converted cruisers, and finally stand-alone capital ships. Carriers had to closely follow the development of naval aircraft as they had to get bigger to accommodate more planes with better technology and longer runway requirements. Only through constant development of naval construction, military aircraft, and carrier operations and doctrine did the *Essex* class fleet carriers emerge from the first experiments with plywood decks appended to gutted hulls.

Path 4 (supply sources) development was also present through the innovations in logistics that would culminated in the Fleet Base Force—the fast fleet oilers and the floating dry docks. Although Path 4 development is more often the discovery or capture of a geographic source of raw materials such as a new oil field or iron ore deposit, here it was the ability for the Battle Fleet to carry its necessary fuel, provisions, and repair supplies with it on maneuvers. The Pacific War needed to end the "tyranny of distance" the fleet would be subject to far from its bases. Therefore, Navy planners designed a substitute— carry the bases with the fleet logistics train. This Fleet Base Force was certainly an amazing sea power development and a very fitting example of a new supply innovation.

The Merchant Marine

Military power isn't the only type of sea power. While naval officers were preparing the sword of American sea power, navalist politicians were leading the charge to develop its commercial arm. The Merchant Marine Act of 1936 focused on the logic transformers of Path 2 development, set the stages for innovation through the Path 5 development of the Maritime Administration, and then left the private sector to respond with new business plans under the new incentives the act offered for Path 3 development. The Maritime Administration served as a government stimulator of private innovation and was not tasked with developing grammar or logic innovations itself, and the stimulation was primarily in the form in subsidies for expanded maritime construction and operation. The rapid expansion of private shipping firms provided the necessary quantity of ships, and lessons learned through the Fleet Problems and other Pacific War requirements helped drive the Maritime Administration's eligibility requirements for the subsidy. The Merchant Marine Act is an example of how government action can stimulate innovation in the commercial sphere. Navalists made great advances in military and commercial sea power, but they could not prevent a negative development in the third point of environmental power, only attempt to mitigate the harm of an error.

A SOFT POWER FAILURE

Undoubtedly, the signing of the Washington Naval Treaty was a mistake in American sea power. The aforementioned problems with the Fortification Clause were a classic example of a sea power heresy—politicians sought to score cheap (and fleeting) political points for disarmament and arms control at the expense of American sea power. Navalists knew that the treaty was a mistake from the beginning. Captain Dudley Knox sought to argue against the treating in his book *The Eclipse of American Sea Power*. While Britain and Japan were both bankrupting their treasuries maintaining large but obsolete fleets, America was in relative ascendance:

> Manifestly the growth of the naval power of Great Britain and Japan was almost stagnant [due to the heavy debt of the two powers], if indeed not retrogressive, and with little prospect of rejuvenation. On the contrary the great building program of the United States, in no jeopardy from lack of funds, was a vigorous, powerful, new growth on the eve of fruition, to which there was no adequate naval reply within the choice of Great Britain and Japan. Even should they have undertaken a similar program at once, ours was too far advanced for them to catch up. Considering the great, highly virile, up-to-date power represented by our program, its rapidly growing condition and the prospect of its early completion, its value as it stood, compared with the relatively obsolete ships of Great Britain and Japan, was much greater ton for ton, probably twice as great.[69]

The Treaty was left in place from 1922 until Japan renounced the agreement in 1936. Once the restriction was lifted, American base expansion on Hawaii, Midway Atoll, and other strategic locations began in earnest. However, through most of the interwar years, navalists had to overcome treaty-inspired disadvantages by expanding the commercial and military forms of sea power to overcome this strategic blunder. The floating dry dock and expanded logistics program was the transformer developed to overcome basing restrictions caused by the politically inspired strategic heresy that was the Washington Naval Treaty.

Lessons Learned

The navalists' actions preparing for the Pacific War offer many important lessons to practitioners of space power attempting to hasten development.

Firstly, naval and merchant marine development efforts offer historical examples of all developmental paths in the General Theory of Space Power. Secondly, the navalists show how the combination of governmental subsidies, proper organizations, and practical exercises can stimulate innovation through these developmental paths. Thirdly, the example of the navalists shows that space power will require the same multidecade effort to adequately prepare us for whatever event—the Spacers War—will become the seminal confrontation to American space power in the future.

The next chapter will use the lessons of the General Theory of Space Power to explore what the Spacers War—American space power's version of the navalists' Pacific War—may be and what actions today's space professionals must take in order to develop space power enough to conquer its challenge in the future. By studying the navalists, proponents of American space power have a ready supply of lessons, cautions, motivations, inspirations, and models with which to innovate today's limited space power into an interplanetary-class arm of American and, by extension, human power to command the future. In many ways, the problems confronting American space power are just as daunting as those that confronted the navalists and will require actions no less heroic than those taken by the great naval officers and politicians of the past. But it is not all an uphill battle. Technology has given us a great edge. As Nofi says in his conclusion on the Fleet Problems and whether they can be recreated today:

> Nevertheless, given the will, we are far better able to conduct serious experiments and maneuvers today than they were during the interwar period, given our superior knowledge of weapons' effects and the power of computers to collect and process data. Free maneuvers should be supplemented by wargaming and the technically sophisticated Modeling and Simulations tools now available. But, as defense analyst James F. Dunnigan has observed, "better M&S tools don't make up for lack of nerve to use them."[70]

Space power can be studied, developed, and mastered using the same techniques as those used by the navalists. We must only have the nerve to use them.

Chapter 5

The Spacers' War—
Beyond Earth Orbit 2053–2057

When America confronts its great space power challenge, what will it look like, and what do we need to do to be ready? This chapter assumes that American space forces are currently in the position that navalists were in 1900. In 2053, American space power, shall we posit, will be called upon to confront a challenge comparable to the Pacific War in intensity and duration in space, with at least some action taking place beyond geosynchronous Earth orbit (GEO), the most highly energetic orbit where the overwhelming majority of militarily relevant satellites currently operate. American space power has some forty years to prepare to confront the challenge. What do we do?

To answer that question, we must first attempt to find out what the 2053 Spacers' War will be. Unlike the navalists, who identified the Pacific as a potential battlefield shortly after the annexation of the Philippines in 1898 and the concurrent rise of Imperial Japan, for the space forces no such potentially looming space crisis is clearly evident. In short, we do not know what the great challenge the Spacers, space power's answer to the American navalists, will be forced to confront. Faced with such uncertainty, what can we do? How can we plan in the face of such uncertainty?

Scenario Planning

Fortunately, strategists (in both military and business fields) have developed a powerful tool for considering an unseen future. The technique is called scenario planning, or "scenarios." In scenario planning, planners consider multiple different futures and develop stories that leaders can use to "test" the validity or appropriateness of projects and programs against each story. Business strategist Peter Schwartz says of the technique:

> Scenarios are a tool for helping us to take a long view in a world of great uncertainty. The name comes from the theatrical term "scenario"—the script for a film or play. Scenarios are stories about the way the world might turn out tomorrow, stories that can help us recognize and adapt to changing aspects of our present environment. They form a method for articulating the different pathways that might exist for you tomorrow, and finding your appropriate movements down each of these possible paths. Scenario planning is about making choices *today* with an understanding of how they might turn out.
>
> In this context the precise definition of "scenario" is a tool for ordering one's perceptions about alternative future environments in which one's decisions might be played out. Alternatively: a set of organized ways for us to dream effectively about our own future. Concretely, they resemble a set of stories, either written out or often spoken. However, these stories are built around carefully constructed "plots" that make the significant elements of the world scene stand out boldly. This approach is more a disciplined way of thinking than a formal methodology.[1]

Dreaming effectively about the future takes shape by forming a set of unique and plausible stories about the future in which the organization will be operating. Each "scenario" tells a story about a way the future may work out. Each scenario must involve actionable details that will directly impact the operations or setting in which the planning organization will act. The scenarios developed are used to test potential future actions (decisions) the organization accomplishing scenario planning are contemplating taking. Scenario expert Kees Van Der Heijden explains:

> These multiple, but equally plausible, futures served the purpose of a test-bed for policies and plans. In Shell [one of the early adopters of scenario planning], an engineering dominated company, most big future-related decisions are project related. Each project is evaluated economically against a set of, say, two or three scenarios, so two or three performance outcomes are generated, one for each scenario. And a decision on whether to go ahead with the project is made on the basis of these multiple possible outcomes, instead of one go/no-go number. The aim is to develop projects that are likely to have positive returns under any of the scenarios. The scenarios as such are not the decision calculus indicating whether or not to go ahead with a project, they are a mechanism for producing information that is relevant to the decision. Decisions are never based on one scenario being more likely than another; project developers optimize simultaneously against a number of different futures which are all considered equally plausible [but not equally probable], and treated with equal weight. In this way both the value and the downward potential of the project are assessed.[2]

A common misconception about scenario planning is that it is meant to accurately forecast the future, that the intent of developing multiple differing futures and planning for all of them is valuable because it allows organizations a better chance of anticipating accurately any future events. Therefore, it is best to develop scenarios that are most likely to happen. This sentiment is wrong. Scenarios are not meant to prepare for a future—they are meant to prepare for any future. They do this not by forecasting, but by preparing the organization to adapt to anything. Van Der Heijden continues:

> Scenarios are not seen as quasi-forecasts but as perception devices. A high/low line approach does not enhance perception, as it does not add new concepts to the "forecasting" frame of mind. Creating three futures along a single dimension, with subjective probabilities attached, is conceptually the same activity as forecasting. It does not cause us to explore conceptually different ways the future could pan out ... *scenarios are a set of structurally different futures.* These are conceived through a process of causal rather than probabilistic thinking, reflecting different interpretations of the phenomenon that drive the underlying structure of the ... environment. Scenarios are used as a means of thinking through strategy against a number of structurally quite different, but plausible future models of the world. Once the set of scenarios has been decided upon they will be treated as equally likely, all being given equal weight whenever strategic decisions are being made.[3]

Thus, considering only scenarios that vary by degree but not by structure is counterproductive. Scenarios must be fundamentally different and, in order to sufficiently expand perception, some must be quite radical and game-changing. Some scenarios may be much like today—indeed, no significant change in structure is a plausible future—but others must present a radically different future in order to test a decision's durability in the face of great stress and transition. In choosing scenarios to develop, there are some guidelines. Schwartz offers, "[s]cenarios often (but not always) seem to fall into three groups: more of the same, but better; worse (decay and depression); and different but better (fundamental change)."[4] These three scenarios present three significantly different futures with which to test decisions. While these are different futures, the likelihood of each one will probably not be the same. However, probabilities should not deter scenario developers. Schwartz says of the role of probability in scenarios:

> In general, avoid assigning probabilities to different scenarios, because of the temptation to consider seriously only the scenario with the highest probability. It may make sense to develop a pair of equally highly probable scenarios, and a pair of potentially high-impact but low-probability "wild card" scenarios. In no case does it make good sense to compare the probability of an event in one scenario against the probability of an event in another scenario, because the two events are assumed to take place in radically different environments, and the assignment of probabilities depends on very different assumptions about the future.[5]

Scenario planning, then, aims to increase an organization's perception of its environment by evaluating decisions or courses of action against a set of fundamentally different plausible (but not equally likely) future scenarios. We can use scenario planning to explore how to prepare for the Spacers' War by developing potential scenarios of how the Spacers' War will begin. We will use the scenarios to test potential technological development efforts (akin to the development of aircraft carriers and advanced maritime logistics platforms) and their suitability for use in each of the possible scenarios.

For our exercise, we will explore four different scenarios which can precipitate the Spacers' War. Each will attempt to be a significantly different future that will fundamentally change the environment with which American space power will have to operate successfully. Each scenario will be plausible but not equally likely to the others, and will highlight a new challenge to national power.

The first scenario will be the "approved" scenario: *Space Pearl Harbor*. In this future, national space power is substantially devoted to integrating into terrestrial power schemes (supporting the warfighter) and the environment remains mostly like it is now. However, a preemptive attack by a terrestrial adversary will test the Spacers' ability to defend and replenish their space assets in time to avert a national defeat.

The second scenario will remain terrestrial in nature but reflect a fundamental shift in American space policy. In *Taking the High Ground*, American space forces are called to take military control of low Earth orbit and enact a police blockade of Earth to prevent adversaries from taking hostile action in space. Adversaries will attempt to lift the blockade and Spacers will be called to defend American hegemony in a space-based *Pax Americana*. This scenario will emphasize a militaristic vision of terrestrially focused American space power.

The last two scenarios will focus on the grammar and logic of planetary defense through potentially unlikely but plausible scenarios in which American space power must radically change to adapt and overcome two of the most threatening scenarios to face human civilization. The third scenario, the *Hammer of God*, will explore the grammar of planetary defense as the cold universe threatens Earth with destruction as a rogue object kilometers wide is on a collision course to strike Earth with a potentially civilization-destroying impact. Only American space power will even have a chance to avert a cataclysm.

The final scenario, *Eat at Joe's*, presents a scenario representing the logic of planetary defense. Here, a potentially hostile intelligence is detected and a ship is on the way. The classic alien invasion scenario is the most unlikely scenario presented and its probability is very remote, but it is still plausible. It may also be the most perception enhancing since it will be the most dangerous and unpredictable. With the fate of the human race in the balance, the Spacers' will be the front line of defense.

These four scenarios will be used to explore decisions on the suitability of potential technologies (enhanced space power elements and idea generation) over all of the scenarios on how the Spacers' War will start. While some foreseeable space technologies will prove very valuable over some scenarios, they may not be as valuable to all of them. Through assessing various technologies through these scenarios, we may be able to find technological paths that the United States can pursue in order to maximize American space power's relevance to whatever situation it finds itself in deep space on the eve of 2053.

Space Pearl Harbor

> As history has shown—whether at Pearl Harbor, the killing of 241 U.S. Marines in their barracks in Lebanon or the attack on the USS *Cole* in Yemen—if the U.S. offers an inviting target, it may well pay the price of attack. With the growing commercial and national security use of space, U.S. assets in space and on the ground offer just such targets. The U.S. is an attractive candidate for a "Space Pearl Harbor."[6]

There is no doubt that the United States derives the most benefits from space power of any nation today. Space power assists in almost all aspects of American activity, be it economic, military, or political. Benefits from its space power may arguably make the United States *dependent* on space activity in order to function correctly. For this reason, adversaries determined to harm or cripple the United States may see American space assets as inviting targets. The Space Pearl Harbor scenario imagines a surprising and crippling attack without warning on American satellites in orbit that demands the Spacers rebuild America's capability as quickly as possible.

America wakes up to a normal day with little news out of the ordinary. Then, while buying coffee on their way to work, people begin to find that their credit cards aren't working. Neither are most ATM cards. Computer internet connections are unreliable. Wall Street fails to open trading due to technical difficulties. Banks cannot access customer records and limit withdrawals. GPS and other satellite navigation systems fail to receive signals and become useless.

At Air Force Space Command, operators lose connection with one GPS satellite, then another, then another, finally most. Communications satellites fail to respond to commands and cease transmitting altogether. Some satellites simply disappear. The problems the country is seeing is due to the satellite infra-structure the economy relies upon is no longer functional. Just how bad losing a large portion of satellites at short notice is to the nation is highly debated with some arguing it would be a catastrophe while others argue it would be an inconvenience and little more. Most agree, however, that we should hope it doesn't happen, and if it did, we must try to recover as quickly as possible.

Let us say the Spacers' War is to replace our space infrastructure as quickly as possible from this attack. What class of technology would be most necessary for rebuilding? Assuming that rebuilding would entail launching new fleets of satellites to repair the ones lost to this event, the overwhelming need would be for advanced launch vehicle technology. It took decades to launch the hun-dreds of satellites that comprise our current space infrastructure, and to replace a large portion of it in a few months or years would require both much quicker launches and far cheaper launch vehicles. Launch vehicles would need to have much higher payload mass capacities as well as be much cheaper to build and launch. Both of these requirements would require new philosophies of launch vehicles to replace those we currently have. Instead of expendable chemical launch vehicles, reusable chemical launch vehicles (assuming they are rapidly recoverable) might need to be used. We might also need to switch to launch vehicles that use nuclear power instead of chemical boosters. We could even scale up our current rockets to produce "big, dumb boosters" that are little more than unsophisticated bottle rockets that do little more than launch extremely large payloads into space itself and place the burden of fine maneu-vering on the new satellite itself. There are many ways launch vehicle technol-ogy could improve to handle the needs of rapidly reconstituting broken space constellations. The key is that the Spacers would need much better launch vehicles than we possess today to compress the time to field entire constella-tions from years to months.

Other technology necessary for this scenario is dependent on just what caused the satellites to fail in the first place. If a large nuclear explosion in space from a hostile actor caused satellite electronics on many of the systems to fail, there may be relatively little debris that could be hazards to space nav-igation beyond the dead, but large and intact, satellites themselves. Even a nat-ural event such as an unfortunate solar flare could produce this type of damage. However, if the Space Pearl Harbor even was caused by attack on satellites from ground-based interceptor missiles or directed-energy weapons in coor-dination with "co-orbital" antisatellite weapons that kill through kinetic means, instead of being simply dead the destroyed satellites may have been shredded, causing enormous amounts of debris to pollute space and make reconstitution far more difficult.

If debris from the event is severe, another important capability in this scenario would be Space Situational Awareness of the near Earth environment (SSA-NE), or what is commonly referred to now as simply "SSA" for lack of significant interest in deep space surveillance among the defense establishment today. The Spacers would need to use ground- and space-based radars and electro-optic sensors to find and track the debris in order to ensure satellites being launched do not hit these dangers and be damaged or destroyed themselves. Robust SSA-NE capabilities would allow mission planners to avoid danger and a great deal of risk by knowing exactly when and where they can launch their new satellites safely. Tracking and cataloguing space debris will also assist with cleanup, another key technology in this scenario.

To date there has been a great deal of talk and little action on developing a capability to remove or otherwise mitigate the danger of debris in space. In this scenario, cleanup capabilities would be a very important third type of critical technology in this scenario. In the event of dead yet intact satellites, cleanup might require little more than attaching a small thruster to the derelict and boosting it into an unused "disposal" orbit and causing it to reenter Earth's atmosphere. In the event of a large cloud of debris, cleanup might require significant advances in technology and massive expenditures deploying cleaning aids such as large electromagnetic "brooms" or laser removal systems. In any case, serious attempts at space debris cleanup will be required to recover from a Space Pearl Harbor event.

A last piece of critical technology for this scenario depends again on the culprit behind the event. A natural event would cause little needed response beyond reconstitution (launch), tracking (SSA-NE), and mitigation (debris removal and system hardening to withstand a similar event in the future). However, if the Space Pearl Harbor was due to an attack, space weapons might be required in order to intercept attacking units before they can strike at the reconstituted American space infrastructure. This leads us to our second scenario.

Taking the High Ground

United States Air Force Air University strategist Everett Dolman argues that the United States should establish a hegemony in space (and thus be in a position to deny any space activity it deems undesirable) in order to both unleash the total human (but most importantly American) potential in space and ensure that the benefits derived from space are used for the betterment of mankind. While other nations may balk at the perceived moral justness of

America being the arbiter of all human space activity, the astropolitik scenario is one that the Spacers may be called upon to pursue. Dolman explains:

> If any one state should dominate space, it ought to be one with a constructive political principle that government should be responsible and responsive to its people, tolerant and accepting of their views, and willing to extend legal and political equality to all. In other words, the United States should seize control of outer space and become the shepherd (or perhaps watchdog) for all those who would venture there, for if any one state must do so, it is the most likely to establish a benign hegemony.

The *Astropolitik* plan could be emplaced quickly and easily, with just three critical steps. First, the United States should declare that it is withdrawing from the current space regime and announce that it is establishing a principle of free-market sovereignty in space. Propaganda touting the prospects of a new golden age of space exploration should be crafted and released, and the economic advantages and spin-off technology from space efforts highlighted, to build popular support for the plan.

Second, by using its current and near-term capacities, the United States should endeavor at once to seize military control of low Earth orbit. From that high ground vantage, near the top of Earth's gravity well, space-based laser or kinetic energy weapons could prevent any other state from deploying assets there, and could most effectively engage and destroy enemy ASAT (anti-satellite) facilities. Other states should still be able to enter space relatively freely for the purpose of engaging in commerce, in keeping with the principles of the new regime. Just as in the sea dominance eras of the Athenians and British before them, the military space forces of the United States would have to create and maintain a safe operating environment (from pirates and other interlopers, perhaps from debris) to enhance trade and exploration. Only those spacecraft that provide advance notice of their mission and flight plan would be permitted in space, however. The military control of low Earth orbit would be for all practical purposes a police blockade of all current spaceports, monitoring and controlling all traffic both in and out.

Third, a national space coordination agency should be established to define, separate, and coordinate the efforts of commercial, civilian, and military space projects. This agency would also define critical needs and deficiencies, eliminate nonproductive overlap, take over the propaganda functions iterated in step one above, and merge the various armed services space programs and policies where practical. It may be determined that in this environment a separate space force, coequal with army, navy, and air forces, be established, but it is not deemed vital at this time. As part of this propaganda effort, manned space efforts will need to be accelerated. This is the one counter to the efficiency argument of the new agency, but it is necessary. Humans in space fire the imagination, cull extraordinary popular support, and, while expensive, Oberg makes the subtle argument that humans "have and will continue to possess a keener ability to sense, evaluate, and adapt to unexpected phenomena than machinery." A complementary commercial space technology agency could be subordinated or separated from the coordination agency, to assist in the development of space exploitation programs at national universities and colleges, fund and guide commercial technology research, and generate wealth maximization and other economic strategies for space resources and manufacturing.

That is all it should take. These three steps would be enough to begin the conceptual transition to an *Astropolitik* regime and ensure that the United States remains at the forefront of space power for the foreseeable future. The details would be sorted out in time, but the strategy clearly meets the elementary requirements previously articulated, from social and cultural to theory and doctrine. It places as guardian of space the most benign state that has ever attempted hegemony over the greater part of the world. It harnesses the natural impulses of states and society to seek out and find the vast riches of space as yet undefined but universally surmised to be out there while providing a revenue-generating reserve for states unable to venture out. It is bold, decisive, guiding, and, at least from the hegemon's point of view, morally just.[7]

Dolman's three steps: declaring a free-market economy and private property in space (essentially abandoning the Outer Space Treaty and its language of space being "for all mankind"), establishing military control of low Earth orbit (in order to prevent from reaching orbit any space activity it doesn't like), and developing a new national agency to establish this new American space effort (essentially developing the Spacers themselves) are relatively simple. The combined impact of all three proposals would be a very welcome acceleration of American space power. However, only the second proposal requires significant technology advancement.

What technology would be necessary to seize military control of low Earth orbit? First and foremost it would require space weapons able to engage targets in Earth orbit and launching into Earth orbit. Weapons that could engage Earth targets as well might also be necessary to ensure a tight and effective blockade. Directed energy and kinetic weapons of a highly selective nature would be necessary. These weapons would also need to be precise and generate little or no debris if used. Most importantly, these weapons would be looking inward. They would target objects beneath them as the weapons would be against Earth originating targets. Space itself is not really an area of interest here. In order to use these weapons effectively, robust SSA-NE capabilities would also be required to ensure hostile spacecraft are found, engaged, and defeated before they can successfully run the blockade. Thus, weapons and SSA-NE necessary to use them effectively are the most necessary technologies for taking the high ground.

However, launch and deep space propulsion will also be critical in this scenario. If a different nation (or coalition of nations) attempts to defeat the blockade, the Spacers will need to have a robust and inexpensive heavy lift launch capability to deploy their weapons before the adversary can challenge them with their own, and be able to reconstitute and sustain their weapons that are destroyed or used up in the event of wartime conditions. Also, Dolman indicates that the entire reason for this military blockade of space is not to establish American dominance per se, but to enforce a regime that benefits all players by providing proper incentives to industrialize the solar system for humanity's benefit. Therefore, large scale deep-space operations to conduct economically useful activity would be encouraged, and the cheaper and more effective deep-space propulsion capability a nation has, the more money it will be able to generate from deep-space activity. Thus, launch and space propulsion is even here a critical technology.

It should be noted that "taking the high ground" need not be rank imperialism from an aggressive United States enabled by warlike Spacers. Indeed, this scenario may be the logical extension from the Spacers successfully recovering from an intentionally hostile Space Pearl Harbor scenario. If the Spacers succeed in rebuilding a crippled American space infrastructure that was deliberately attacked, it would be in America's (and probably the world's) interests to never let a sneak attack in space like that ever happen again. It may well be that a military blockade of low Earth orbit will arise from a stark lesson in the danger to the world of space pirates pulling off a Space Pearl Harbor.

Another potential offshoot of this scenario is that the Spacers may actually be called upon to *prevent* an attempt by a different nation to seal a military blockade of low Earth orbit. If the Spacers resist the establishment of a foreign blockade, they will also need the same technology as the blockader, with the minor alteration that the weapons they deploy will need to look up as much as look down. Whether or not the Spacers confront this scenario as blockaders or blockade runners, high technology will be necessary and this scenario need not arise from strict American aggression. Taking the high ground is a scenario that the Spacers may need to confront, and this scenario does not automatically turn the Spacers into the "bad guys" either. In fact, this scenario might give birth to the technology that might ultimately save the human race and Earth itself.

The Grammar of Planetary Defense: The Hammer of God

Issues of planetary defense—defending Earth and its inhabitants from threats arising from space itself—comprise the last, and most remote, two scenarios. This third scenario begins with humanity discovering that a large rock or comet that is estimated to impact Earth in a number of years. The only way to prevent massive damage and possible extinction of the human race and most other life is for humanity's space power to destroy the impactor or alter its course to avoid hitting Earth. This scenario is called the grammar of planetary defense because there is no malevolent intelligence at work against the human race or an enemy to defeat. This problem arises solely through the operations of the mechanical universe. Thus, it is a grammar problem—humanity's capability to control nature is being tested, not its ability to defeat an active foe.

Planetary defense against the impact of dangerous space bodies has been the subject of much interest in NASA and some interest in the Department of Defense. Indeed, the term "planetary defense" in government circles focuses exclusively on this grammar scenario, generally because the grammar scenario has gained some traction in serious circles, while a logic scenario has no real chance of eliminating the "giggle factor." But there are important differences between NASA and the DOD. While NASA tends to think of planetary defense in terms of a technology problem, the DOD tends to think programmatically. Air Force Lieutenant Colonel Peter Garretson laments the lack of intensive thinking in military circles:

> Planetary defense may seem an abstract and unreal national security risk. However, it proved quite a serious problem for the dinosaurs, who previously inhabited our planet, and it poses no less a threat today. No matter how remote some people might think the chances of having rocks fall on their heads, they should at least be concerned that no government or DOD contingency plan exists to counter an impact or mitigate its consequences.[8]

Even without a conscious planetary defense strategy in the DOD, the military has nonetheless provided a large portion of our current planetary defense capability, especially in the area of detection through its space situational awareness capabilities:

> As part of the Strategic Defense Initiative (often called Star Wars), the U.S. Department of Defense spent tens of millions of dollars in developing an electronic sensor that downloaded its data rapidly into computer memory. This innovation got around one of the major drawbacks of previous systems, and meant that a new image could be exposed while the previous image was being stored, allowing for essentially continuous data collection.
>
> At the forefront of this work were the Lincoln Laboratories of the Massachusetts Institute of Technology. At that institution, others were involved in satellite tracking and the like using an observatory complex near Socorro, New Mexico. That site included not only operational GEODSS (Ground-based Electro-Optical Deep Space Surveillance) cameras, but also two identical systems that were rarely used. The leader of the Lincoln Labs team, Grant Stokes, arranged for a series of test observations to be made with one of those cameras, starting in 1996 using a conventional detector system. As expected, their results paralleled that of the NEAT

(Near Earth Asteroid Tracking) team, who had by then started their work on Maui. By the middle of 1997, a handful of new NEOs (near Earth objects) had been discovered from the New Mexico site, along with dozens of other asteroids. Stokes reckoned that it would be a simple matter to scale up to the larger, more efficient detector that the military had developed. With that plan in place, he reasoned, their NEO discovery rate should rocket. And so the Lincoln Near-Earth Asteroid Research (LINEAR) project was born....

LINEAR's output has been phenomenal. In the first two years of operation, almost three million asteroid observations have been supplied to the central data repository. Four hundred thousand asteroids were detected; 60,000 of these were recorded in sufficient detail to be given new designations (such as 2005 JG5).... The significant point is that the LINEAR team has discovered (as of 2000) more than 400 NEOs. Previously, the global discovery rate was peaking at between five and ten per month. Now LINEAR alone is finding Earth-approachers at more than one per night.[9]

As of 2011, LINEAR has discovered 2,423 NEOs and 279 comets.[10]

Although the DOD provides a monumental contribution to our current space detection capability, NASA is still considered by most to be the "point man" for planetary defense activities. However, a solid argument can be made to give this mission to the DOD:

> Since no U.S.-assigned or -authorized planetary-defense missions exist, the DOD, as an organization, does not have any "impact defense" operations. Few individuals in the DOD perceive this lack of policy as a problem, and those few who do must contend with the giggle factor. This train of thought suppresses any further acknowledgement or research. Assignment of responsibility would rectify this problem, yet who should assume responsibility for a planetary-defense mission? Readers might wonder why the authors mentioned STRATCOM (United States Strategic Command) as a possibility. Why not some other part of the DOD? Why the DOD at all? Perhaps NASA could handle detection, reconnaissance, and mitigation missions while trying to replace the space shuttle and return to the moon. Maybe the DHS or Federal Emergency Management Agency (FEMA) represent a better option since impacts might become a national disaster.
>
> Both NASA and the DOD have expertise in space matters and operate space assets, but NASA's core mission is space exploration. The DOD's core missions are maintaining U.S. security, protecting American lives, and ensuring the security of our allies. Expertise aside, planetary defense is clearly a defense mission. Further, since the DOD maintains a robust space mission, the proposed mission appears more closely aligned with the strengths and scope of the DOD than with those of the DHS.
>
> Within the DOD, possible options might include AFSPC, the National Security Space Office, the Missile Defense Agency, and STRATCOM. Several reasons make STRATCOM the best option. For one, STRATCOM's mission calls for "provid[ing] the nation with global deterrence capabilities and synchronized DOD effects to combat adversary weapons of mass destruction worldwide." The com-

mand coordinates DOD capabilities to thwart weapons of mass destruction. We can consider an inbound Earth-impacting rock a weapon, despite the absence of an adversary. A combatant command, STRATCOM has the established lines of communication and the authority to react to strategic-level threats. It already maintains global vigilance and space situational awareness. The former U.S. Space Command has been dissolved and subsumed by STRATCOM. Through AFSPC, the command already maintains daily space surveillance for detecting launches of ballistic missiles and tracking artificial satellites and Earth-orbital debris. Although AFSPC maintains space assets, operational control falls under STRATCOM's authority. It also controls all military nuclear capability, perhaps the only option in certain minimum-warning scenarios. Moreover, STRATCOM is well practiced and competent with respect to disseminating rapid warnings to civilian leadership and civil defense networks. Finally, the command has years of experience in negotiating and executing collective security arrangements, such as that of the North American Aerospace Defense Command with Canada and those involving the North Atlantic Treaty Organization[11]

The Spacers will be children of both NASA and the DOD, so who is ultimately made responsible for planetary defense today is nowhere near as important as developing the capabilities necessary for effective planetary defense. In this scenario, detecting and categorizing planetary defense threats are the most important concerns. Thus, the ability to monitor deep space is the single most important capability to have in this scenario. We'll call this capability space situational awareness for deep space (SSA-DS). An effective SSA-DS capability should be built upon our current capabilities such as LINEAR and other systems, but must include space-based augmentation such as outward-looking telescopes placed in the inner solar system (i.e., Venus-like orbits) such as missions proposed by the B612 Foundation. Seeing a threat coming with as much notice as possible is the lynchpin of the grammar of planetary defense campaign.

After adequate SSA-DS technology is developed, space propulsion for both heavy lift and deep space propulsion will be necessary to combat the threat. With sufficient warning, deep space propulsion may be the last technology necessary to successfully defend against this scenario. With many years of lead-time, attaching a continuously operating thruster (such as a mass driver which uses the object's material itself as propellant, the larger the better) may be able to "nudge" a threat out of the way without any additional operations necessary. Thus, mature SSA-DS and space propulsion technology may be all that is required. However, even this relatively benign technology requirement causes hesitation among some. Author Duncan Steel argues:

> There is another aspect to [planetary defense], though. Suppose that we—mankind—develop a defense system capable of diverting asteroids away from the

Earth. That sounds great, even essential to our long-term survival, if you agree with this book so far. But the downside is this: if a nation has the ability to divert NEOs away from the Earth, then it also has the ability to deflect them towards the planet.

Would that be suicide, you might ask? Not necessarily. Imagine that nation A spots a hundred-yard asteroid that will just miss the Earth. Without telling anyone, it could divert the projectile, while still far away, in such a way that it will slam into enemy country B, wiping out its capital city with a hundred-megaton explosion. And nation A could claim no prior knowledge. Or country X might see another small asteroid heading for its own territory and, in self-defense, divert it in such a way that it hits nation Y, a whole ocean away.

This scenario is called the deflection dilemma. The ability to protect ourselves from impacts by NEOs is a double-edged sword. Because of this, many civilian scientists, including this author, argue that there is no need to build asteroid defense systems until an actual threat is identified. We should carry out an appropriate benign surveillance program first.[12]

Steel's last paragraph is extremely disturbing and should call into question any scientist or NASA-led planetary defense program. Steel and his "many civilian scientists" display a remarkable level of naïveté. Preventing the development of potentially Earth-saving capabilities on the off chance someone can use it as a weapon is patently ridiculous because it would essentially prevent anything from ever being developed! Any technology runs some risk if it falls into the wrong hands. A more pragmatic argument against this "deflection dilemma" is simple reality. Lieutenant Colonel Garretson argues:

> Having a decade of advance warning might seem like plenty of time to construct these policies and a mitigation operation, but it isn't. We would need most of this time to slowly affect the velocity of an asteroid with a low-thrust, high-efficiency tug. Reaching a menacing asteroid will take several years of flight time as well. Clearly, we need mission planning, spacecraft development, and testing. Current Department of Defense (DOD) system development and procurement can easily run longer than a decade. The F-22 fighter aircraft alone has taken nearly 25 years to evolve from a list of requirements to initial operating capability.[13]

Thus, even if we are lucky enough to have significant warning time and only require the ability to nudge a threat out of the way in order to mitigate its danger, we need to prepare now. The "deflection dilemma" is no dilemma at all, just a fever dream of paranoid minds with an ax to grind against the military. The Spacers must be against any such irrational thinking, especially since if advance detection fails and our SSA-DS capability is insufficient to provide advanced warning, the Spacers will then require the most powerful and destructive space weapons we can imagine to destroy, instead of divert, the threat. Heavy and destructive weapons (probably nuclear) deployable in space along with the techniques and capability to launch many in a coordinated campaign to virtually annihilate an incoming threat may be necessary to survive a planetary defense scenario. In this situation, squeamishness and moral preening against "space weaponization" will be a direct threat to life itself and any instances of it must be mercilessly crushed in debate and promoters subjected to the most wilting criticism possible. Engineer and writer Travis Taylor makes a blunt but accurate assessment of space weapons for planetary defense:

> And how about one day when we finally detect that asteroid, or comet, or near Earth object of some sort that is going to slam into the Earth and cause major problems for us? We will be scrambling to figure out how to put the right types of weapons into space in a very short time frame that may have some impact on our impending doom. It would be a lot simpler if we already had a design for a platform, just completed, prototypes to experiment with, and at least one or two systems deployed in orbit. Perhaps something that's not just in orbit, but on the Moon, or at a Lagrange point or somewhere else in space. But the point is, we need the planetary defense systems, weapons systems, in space before it's too late and we're invaded by a world killing object such as the one that gave the dinosaurs such a problem.[14]

If heavy weapons are necessary to adequately defend the Earth, space propulsion will again be vastly important to make sure any systems we develop will be adequately deployable to accomplish their mission. The combination of adequate SSA-DS capabilities, heavy space weapons, and adequate heavy lift and deep-space propulsion technologies both make possible and require a large-scale human presence in space. Journalist William Burrows makes the argument that planetary defense is ultimate expression of large-scale space power:

> There has to be a true, continuous presence [in space]; a presence that has a compelling purpose. And the continuous presence will, in turn, require much less expensive launch systems than are now in use. Yet however relatively inexpensive the launch vehicles and spacecraft that carry people and cargo to the Moon, the inescapable fact is that going there will be expensive indeed. But there is no alternative except what could be the ultimate catastrophe.
>
> The most daunting obstacle to a permanent program to use space for the protection of Earth is not financial or technical. It is political. It is of utmost importance that a bipartisan planetary-defense culture takes hold in the United States and around the world and accepts that space budgets must not only grow, but must be stable and protected over the infinitely long term, rather than be debated and redebated every year. Planetary defense must, in other words, be as normative as the military. No government would consider abandoning its armed forces. Protecting Earth, as its constituent nations are protected, should become permanently institutionalized and financed accordingly. There is a model. The navy budgets the operation of large fleets, such as aircraft carriers, for the expected life of a ship. It is inconceivable that a $4 billion supercarrier, which takes seven years to construct, would not have enough operating funds so it could fulfill its mission over its

projected lifetime. As the carrier admirals do not have to scratch for funding to operate their ships every year, neither should the managers of the spacecraft fleet, the stations, and the lunar colony. Their funding must be as steady and dependable as the military's.[15]

The bipartisan planetary defense culture is necessary to grow, mature, and support the Spacers and their mission to defend the Earth as well as American space interests. Indeed, the managers of the spacecraft fleet, the stations, and the lunar colony Burroughs believes is necessary for a mature and dependable planetary defense capability will be the Spacers themselves, and the Spacers will be far more valuable than simply defending against rogue space bodies that threaten Earth. There may be even greater dangers lurking in the depths of space.

The Logic of Planetary Defense: Eat at Joe's

Aliens emerging from the depths of space to kill, eat, or otherwise inconvenience Earth is the classic science fiction cliché. However, it is also a plausible event and certainly the one scenario that will provide the ultimate challenge to American—indeed, human—space power. It is the logic of planetary defense because the Spacers will not be fighting against nature—they will be defending against a thinking and responsive enemy. Unlike the soulless impactor from the grammar scenario, the enemy "has a vote" in which side will win.

Though people who think about alien invasions are most often science fiction and fantasy writers, the subject has not completely escaped modern military strategists. Esteemed scholar and defense strategist Dr. Colin Gray in his writing on space power has addressed the issue:

> Notwithstanding the vast asymmetry between the terrestrial geographical environments and space, it is not entirely obvious that "the stars" or "the heavens" have strategic significance for contemporary defense planners. Threats originating from far beyond the Earth-Moon system may appear from beyond our solar system or even from beyond our galaxy. If they do, we will be fortunate if we are able even to note the approach of such threats, let alone be equipped to see them at launch. In the long run, the very long run indeed, the security of the human race most likely will depend upon its space power. The dinosaurs faced a grim prospect between emigration and extinction and were condemned technologically to the latter. Fortunately for us, the random menace from fast-moving alien objects in space would appear to pose far more severe a threat to life on Earth than does purposeful menace from alien civilizations that would be unschooled in the niceties of the Geneva Convention. An asteroid may just terminate the human experience and settle religious arguments, but at least in principle it is detectable, trackable,

and possibly divertable. By way of caveat, any animate, purposeful, alien menace that could reach Earth from another solar system, let alone from another galaxy, can be assumed to be likely to enjoy a decisive technological edge for superior strategic effect.[16]

Let us assume for this scenario that Earth got a bit luckier than Dr. Gray's scenario. Due to an incredible coincidence, random luck, or divine intervention, Earth-based observers have identified an immense object emerging from the Kuiper Belt into our inner solar system and projected to travel through the asteroid belt into the rocky planets. It is big (many kilometers across and many more deep) and dark. The object is not attempting to communicate on any band that we can detect. Indeed, it seems as if the object was designed to specifically limit radiation of any kind and conceal itself as much as possible. Like an attack submarine, this alien *Nemesis* craft is running silent. Worst, it just looks *evil*—as if it emerged whole from a Lovecraftian nightmare. Perhaps our fears are just an illogical expression of our fear of the unknown and bias toward our terrestrial aesthetics. Perhaps not. In any case, the Spacers are forced into becoming our first line of reconnaissance and, God be with us, our first line of defense against an attack of potentially limitless power.

Pop television scientists, especially on the obligatory alien episode of various documentary science television shows, tend to dismiss talk of defense against alien attack by saying that any aliens with the capability for interstellar flight would be so advanced that any resistance by humanity would be useless and, hence, not worth discussing. This is an unsatisfactory response since it both dismisses the impact of other potentially mitigating factors significant in military analysis and real professionals would be derelict in their duty to those they defend if they help such a flippant and defeatist attitude. For instance, traveling interstellar distances to reach Earth with an invasion force is evidence of superior technology, but it also may reveal a critical enemy vulnerability—astronomically large lines of communication. One of the most critical pieces of intelligence in an alien invasion scenario is to understand how the adversary got here in the first place. Knowing their method of interstellar travel is essential, because perhaps the one advantage humanity would have in such a situation (and one potentially powerful enough to negate a technological advantage) is the logistical advantage. An alien attack fleet, if it needed to cross the interstellar void, will probably not be able to get reinforcements easily. Conversely, the entire strength of the human race will be available to the fight entirely in the area of operations (our solar system). If alien interstellar travel is "hard," i.e., it is slow (sub–light speed) or mass-limited (the mother ship is running on fumes and is the bulk of what we're likely to fight), certain military targets and strategies can be devised.

But adversary technology is not the only question we must concern ourselves with. The space power ability possessed by the Spacers will be equally critical to determining how humanity should defend itself. Obviously, the more technology humanity had (space power–centric or not) the better. However, the type of space technology we have will in large part limit our strategic alternatives. Michael Michaud provides us with a very interesting dilemma about alien colonization and planetary defense with which to stimulate our thinking:

> Tipler and Barrow, who visualized an aggressive interstellar expansion and colonization program, attempted to draw a distinction between that idea and the more pejorative concept of interstellar imperialism. First, they declared that there was no reason to expect imperialism. Then they acknowledged that the existence of "imperialists" would motivate "colonizers" to speed up their occupation of previously unoccupied solar systems, in order to prevent the "imperialists" from seizing them. They cited the rapid conquest of central Africa by European powers as an example of such behavior (African territories were, of course, already inhabited).
> Tipler and Barrow seemed not to recognize that a contest between "colonizers" and "imperialists" would be a contest between empires. Nor did they admit that the arrival of a probe from another civilization might be seen as threatening. They even claimed that the colonization by extraterrestrials of all the planets in our solar system other than the Earth would not be imperialism, because the planets are just "dead rocks and gas." Yet, they admitted that the alien colonization of uninhabited planets would prevent the native intelligent species from eventually colonizing these worlds. Imagine what our reaction would be if we saw extraterrestrials colonizing Mars.[17]

How would we react if we saw aliens colonizing Mars? Let us envision two different classes of answer to that question by defining two different models of planetary defense: the Aristotelian view and the Copernican view. The Aristotelian view of planetary defense is "Earth centric." The Aristotelian answer, then, to Michaud's question is that we would do nothing if we saw extraterrestrials colonizing Mars, or at least we would not try to stop them. To the Aristotelian, Earth is human real estate. Everything else is negotiable. Some advantages of an Aristotelian view would be that there would be less risk of conflict because humanity's perceived sphere of interest would be limited to the Earth, or perhaps Earth-Moon system. If *Nemesis* stopped at Jupiter or Mars, problem solved! In the event of an attack on Earth, human forces would also be concentrated on Earth rather than potentially spread over the solar system, enhancing the military principle of mass. Some disadvantages of this planetary defense scheme include leaving Earth more vulnerable to attack by letting the enemy closer to "our homes and firesides." Also, as noted by Michaud, alien colonization of other worlds of our solar system would prevent us from potentially doing the same.

Alternatively, the Copernican view of planetary defense is "Sun centric." The Sun is the center of the solar system; humanity is the native intelligent life in the solar system; therefore, the solar system belongs to humanity. The Copernican would consider any extraterrestrial colonization of our solar system without human consent to be an event worthy of armed resistance. Thus, if we saw extraterrestrials colonizing Mars, the Copernican would launch a military fleet to cause the enemy to cease, come to some mutual agreement, or physically prevent the adversary from completing its colonization efforts.

A Copernican view isn't inherently imperialist in and of itself. Native life might exist throughout the solar system in the many suspected liquid oceans of Europa and other Plutoid worlds or other environments more exotic. Indeed, other intelligent life native to our solar system may exist. A Copernican could easily accede to general rules declaring that where native life exists, that area rightfully belongs to its native species. However, it does appear that humanity is the only technic species native to our solar system. Indeed, a Copernican crusade against an extraterrestrial colonizer might even be in the defense of an extraterrestrial, but solar native, life-form being threatened by the non-native colonizers.

A large factor in whether we adopt an Aristotelian or Copernican view of planetary defense is the current level of technology and ability to operate in space. In order to prosecute a Copernican strategy, we would need to be able to contest space beyond Earth. That requires access to space with large payloads to prosecute military operations—a great deal more space power than the United States now possesses. In 2014 we simply could not do much more than watch *Nemesis* colonize Mars and wait for a friendly message of "Hello, neighbor!" while praying against an attack on Earth after the alien beachhead is complete. An Aristotelian view of planetary defense is the default position of a space power which does not have the ability to perform large logistics operations throughout the solar system. A Copernican planetary defense strategy is open only to those space powers that have widespread access to their solar system. Any decision on which planetary defense strategy to prosecute should be based on the needs of the human race and on considered and reasoned strategy chosen for its merits. The decision should not be made by default due to a limited technological situation. Thus, space propulsion is a vital technological need of this scenario.

But space propulsion is but one of the technologies that the Spacers would need in this scenario. The most obvious and important would be heavy space-capable weapons. Big ones. As big as can be produced. Without space weaponry that can look out (as opposed to the Earth-facing kind important to the Hegemon scenario), the Spacers can put up no defense against an invasion. The invasion would need to descend into Earth before any defensive fire could be attempted. Nuclear and more exotic directed-energy weapons would represent the *sine qua non* of any space-based defense. In addition to heavy space weapons will be the targeting equipment that will allow them to be used. Space situational awareness for deep space (SSA-DS) will be critical complementary technology for the Spacers deep-space weapons both for detection and targeting. Command and control technology that will seamlessly integrate both weapons and SSA-DS must also be deployed.

When weapons, detection, and C2 are advanced enough to enact militarily useful strategies and mount a potentially successful defense, the limiting factor will be space propulsion. Both high payload launch technologies and deep-space propulsion will be needed to deploy and logistically support any weapons developed, and the capabilities of propulsion technology will be the critical factor in how far out an invasion force can be engaged from Earth and whether the Spacers can adopt a Copernican strategy or be forced into an Aristotelian defense. Thus, even though a fighting capability in space is a necessary condition for defense, as soon as an effective capability is reached its utility will be defined primarily by the space propulsion capability available to deploy and sustain it. Even though space propulsion is not the most important capability for a logic of planetary defense scenario, it is still vastly important because it may offer the most strategic freedom of any technology.

The Unity of Planetary Defense

Just as space power cannot be completely appreciated without an understanding of the symbiosis between space power grammar and logic, so does planetary defense posses an essential unity. Planetary defense is planetary defense, regardless of the threat that it must contest against. Therefore, planetary defense is the most robust of the scenarios presented. It is also probably the issue with which the Spacers will ultimately prove their worth the most. Preparing for planetary defense, focusing on either the grammar or logic scenario, causes the Spacer to consider all aspects of space power and drives the need for the most robust technological and industrial base possible. Michael Michaud argues:

> There is still no concrete plan in place for humanity's response if we discover an asteroid heading our way. Some analyses indicate that we can divert Earth-crossing asteroids and comets only if we reach them years before their projected impact. For an asteroid 200 meters in diameter, we would need roughly 20 years; for a larger asteroid, the lead time would be longer.

To starflight theorist Gregory Matloff, that meant building an infrastructure in the outer solar system—at a minimum, the lookout posts that would watch for incoming bodies. That capability also could give us the means for spotting other potentially dangerous intruders—the probes or inhabited vehicles of another civilization. Earth security would be extended to solar system security.

Planetary defense can be seen as a rehearsal for direct contact. It provides one model of preparing ourselves to deal with the exploring machines of a more advanced technology. Whether we could defend ourselves would depend on the relative capabilities of the two civilizations. Whether we would need to would depend on the intentions of the more powerful one.

SETI [Search for Extraterrestrial Intelligence] conventional wisdom assumes that because we will be much less technologically advanced than any other civilization that we contact, we would be helpless if the extraterrestrials were hostile. This disparity may turn out to be true, but it remains unproven. To assume our weakness in advance would be preemptive capitulation.[18]

Preemptive capitulation will not be in the Spacers' vocabulary. But the essential unity of planetary defense also means that preparing for one will automatically provide benefits for defending against the other. Indeed, requirements for new technologies can be developed and new scientific discoveries can be generated by taking even the most basic first steps to building planetary defense capabilities. Even better, planetary defense technologies will even provide immense benefits to our existing exploration initiatives. Dr. Michael Papagiannis provides a very interesting rationale for a much more enthusiastic and intensive solar system exploration program. According to the Harvard-trained and Boston University physicist, current limitations in our SSA-DS capabilities prevent us from even being certain that we are indeed the only intelligent species currently inhabiting our solar system:

A small number of space colonies ... 1–10km in size, could have easily escaped optical detection from Earth, lost among the multitude of physical objects in the asteroid belt....

We must undertake, therefore, a concerted search for extraterrestrial activity in our own solar system and especially in the asteroid belt. This investigation must include fly-by missions to certain selected asteroids, as well as observations from the Earth using both ground and space-borne telescopes covering many different spectral regions. The intent would be to search for asteroids with peculiar physical properties such as unexpectedly high infrared temperatures, for manifestations of advanced civilizations such as radio transmissions, and for any evidence of technological activity such as by-products of nuclear fusion.

The likelihood of finding extraterrestrial colonies in the asteroid belt is probably very low, but it is certainly worth undertaking because the pay-offs could be very high. Even if we were to find nothing, this should not be considered as a failure of our efforts, because exciting as it might be to find other civilizations, it is equally important to know that our solar system and hence the galaxy have not yet been colonized and that therefore we are probably alone in the entire galaxy. Also, and independently of the search for extraterrestrial life, the new knowledge on the origin and evolution of our solar system to be gained from a direct exploration of the asteroid belt would certainly make the effort worthwhile.[19]

Thus, an exploration program based on the needs of planetary defense that include cataloging potentially dangerous space objects that may threaten Earth as well as looking for space objects with physical properties that may indicate artificiality would offer scientific benefits exceeding that of our current exploration program. Such efforts may be even more sustainable than our currently exploration program due to the addition of very pragmatic reasons to explore beyond simple curiosity.

Even the technologies necessary for a robust planetary defense capability would have broad application to any activities we decide to pursue in space, like the space power theory in Chapter 1 suggests. For instance, Travis Taylor offers the following list of technologies with value to counter an alien invasion:

> The following are examples of ... technologies that should be investigated [to prepare for an alien invasion]:
> - Railgun technology
> - Space Launch Earth-to-orbit technologies
> - National Missile Defense
> - Tactical, ground, air, and space-based directed-energy weapons
> - New types of armor (powered armor and mecha)
> - EMI [electromagnetic interference] weapons
> - Micro, nano, and pico satellites
> - Nanotechnology
> - Advanced software weapons
> - Directed solar energy
> - New energy sources
> - Defensive shields
> - Ultra-large arrays for optics and RF [radio frequencies]
> - Lunar catapults
> - Space-based missile platforms
> - Space, lunar, Mars, etc. bases
> - Space navy
> - Civil defense shelters
> - Others yet to be discovered.[20]

This list has many different technologies, all with utility beyond the planetary defense mission, and many can be developed (at least in part) through terrestrial research and development. Therefore, not all of these technologies must rely on the Spacers, or any space power agent, to necessarily develop them. In fact, the one technology Taylor says is most important to planetary defense is the one technology that underpins all others in space technology:

> What is the most serious deficiency in humanity's planetary defense capabilities? The answer is obvious: space propulsion. At present, no one on Earth can conduct even the most limited military operations in space, even inside the Earth-Moon system. The reason is that no remotely adequate propulsion systems exist. At least as far as the public knows, no one is making any serious effort to rectify the situation.
>
> If the solar system were invaded, humanity would be in the unenviable position of being forced to defend our home planet on that planet. History provides us with many examples of nations that were forced, for one reason or another, to fight defensive wars entirely within their own borders. Sometimes they were successful, sometimes not, but they invariably suffered heavy civilian casualties and heavy economic losses.
>
> On the other hand, we have the example of England. No enemy has landed on England's shores in more than nine hundred years, largely because of her naval superiority. Of course, naval superiority will not protect you from invasion if you have enemies on your borders—but a planet is, like Britain, an island.[21]

Thus, it would seem that space propulsion is a common theme for all of the scenarios presented that might emerge as the Spacers' War. In fact, space propulsion underwrites all space power activities, and increased capabilities in space propulsion makes everything else in space easier. Space propulsion truly "lifts all boats" in planetary defense, and in space power.

Space Strategy through the Scenarios

Our explorations of the four scenarios that might describe the Spacers' War of 2053, ranging from a great power astropolitcal struggle to planetary defense against interplanetary debris or alien intelligences, have often stressed the limit of credulity. From a strict probabilistic mentality, each scenario may not be likely and may even be completely dismissible for the majority of people, but all of these scenarios are plausible—in that none of them can be dismissed completely. To judge the validity of this exercise, we must not forget that scenario planning is not meant to forecast the future, but to expand one's thinking:

> To operate in an uncertain world, people need to be able to *reperceive*—to question their assumptions about the way the world works, so that they could see the world more clearly. The purpose of scenarios is to help yourself change your view of reality—to match it up more closely with reality as it is, and reality as it is going to be.... *The end result, however, is not an accurate picture of tomorrow, but better decisions about the future.*[22]

Albert SzentGyorgi tells a story of a group of soldiers lost in the mountains (relayed by Weick):

> A small Hungarian detachment was on military maneuvers in the Alps. Their young lieutenant sent a reconnaissance unit out into the wilderness just as it began to snow. It snowed for two days, and the unit did not return. The lieutenant feared that he had dispatched his men to their death, but the third day the unit returned. Where had they been? How had they found their way? "Yes," they said, "we considered ourselves lost and waited for the end, but then one of us found a map in his pocket. That calmed us down. We pitched camp, lasted out the snow storm, and then with the map found our bearings. And here we are." The lieutenant took a good look at the map and discovered, to his astonishment, that it was a map of the Pyrenees.
> Weick suggests: "If you are lost any old map is better than nothing." The map enabled the soldiers to get into action. They had been mentally disabled; but now the map, believed to represent the surroundings, gave them a new feeling of understanding and a reason to act.... The map got them out of the paralyzed state that they were in. Accuracy did not come into it.[23]

So, from what we have learned through this book and these scenarios, let's develop a map. None of our scenarios were developed enough to provide an exact description of events that may challenge American space power in the future, but what we have learned generally can provide us with a map with which to prepare for the unknown future. Our map will consist of three lines of effort that will prepare the Spacers for whatever war they might confront: a technological line of effort, and operational line of effort, and an organizational line of effort. These three efforts will not be a complete plan of space power development but will provide a framework with which to develop a complete space power program to develop the space power program America needs to be ready for whatever the future may throw at her.

Technological Line of Effort: The Nuclear Hammer

All of the scenarios briefly reviewed in this chapter had one beneficial technology in common—that of space propulsion. In order to confront any of the four scenarios, either launch vehicle or deep space propulsion technology (or both) would have to be remarkably improved. Fortunately, we have a prime contender in the new technological leap that would provide American space power the propulsion revolution that true space power requires: nuclear energy.

Recalling our S-curve discussion in Chapter 3, current launch vehicle technology using chemical rockets can be considered at the very top of its

S-curve. That is, even with significant effort expended on improving chemical launch vehicles, there is no more improvement in capabilities to be had. A common lament among space enthusiasts is that we are using the same launch vehicles today as we were using in the 1960s. This is false. Today's Atlas V and Titan IV launch vehicles are far more sophisticated than their earlier Atlas and Titan forefathers, and they have infinitely better technology. Unfortunately, due to the inherent limitations of chemical rocketry, the improvements made to or current generation of launch vehicles do not significantly improve the efficiency and effectiveness of our rockets. Chemical rockets simply cannot get much better than they are now. New improvements such as flyback and reusable boosters can advance the state of the art somewhat, but even with these advances chemical rockets can only take us so far in space.

What is necessary is to make a leap to another rocket power source that promises a new S-curve that can ultimately extend space propulsion capabilities to levels that will open the entire solar system and beyond to American space power. Nuclear rockets in the form of nuclear thermal rockets (or NTRs, essentially conventional rockets that use nuclear rather than chemical reactions to generate heat for the exhaust propellant) and nuclear pulse propulsion (or NPPs, rockets that use the external detonation of a small nuclear device to propel the spacecraft to incredible speeds at incredible efficiencies) promise easy transitions from chemical rockets as well as the potential to offer an S-curve that can ultimately take us to easy interplanetary travel and offer us an entry-level capability for interstellar travel.

The United States is already reasonably far along the nuclear space propulsion S-curve. Prototypes of NTR engines with demonstrated efficiencies and lift capabilities beyond conventional chemical rockets were already tested extensively in the 1960s under the NERVA (Nuclear Energy for Rocket Vehicle Applications) Program, indicating that we are far enough along the S-curve that an operational jump from chemical to nuclear launch vehicles will already provide a capabilities boost. There are some engineering problems that will still need to be addressed in order to develop a proper launch vehicle, such as improved protections against radiation leakage and other safety issues, but the engineering challenges are relatively well known for a first-generation operational vehicle. NTR applications for deep-space propulsion will also provide great leaps in American capabilities. But a basic NERVA-type NTR vehicle is only the first draft of a technology that promises much greater effectiveness later.

Steam-powered oceangoing vessels began in earnest with the cross–Atlantic voyage of the SS *Savannah* in 1819. Its long and dangerous crossing at only a handful of knots is a far cry from the gas turbine–powered speed and safety of a modern U.S. Navy destroyer. However, technologies used in both ships are remarkably similar to each other, and the modern destroyer can be considered a product of "evolutionary innovations" that took the steam power S-curve from the humble beginnings of *Savannah* up to the present day. While modern ships were perhaps far beyond the technological prowess of the men who built the first steamships, the designers of *Savannah* would still be able to see much of *Savannah's* plant in today's marine engines. So can we imagine that much more capable spaceships will be able to be built in the future using the same basic principles behind the NERVA NTR concept. The two basic limitations of an NTR rocket are the molecular weight of the propellant (ideally liquid H2) and the heat of the nuclear reaction. The NERVA NTR is based on solid core (i.e., the reactor stays in a solid state throughout operation) and its temperature performance is limited. Thus, a nuclear reactor that remains solid only provides a little performance increase over chemical rockets.

However, the NTR concept, as James Dewar states, "is not constrained by heat and weight, like the chemical rocket engine, but by the state of knowledge at any given time."[24] Specifically, how to contain high temperature reactions. A solid core NTR can achieve specific impulses of about 800–1250 seconds with high thrust and weight capacities. However, as materials handling and temperature containment technologies improve (which are evolutionary in nature, not necessitating jumping to new S-curves), liquid and gas core NTRs will likely become technically feasible. With this technology, vastly higher reaction temperatures are possible, leading to specific impulses in the multiple thousands, perhaps as high as 8,000 seconds. With NTR engines reaching theoretical limits (and hence, S-curves that reach) this high, the NTR portion of the nuclear space propulsion S-curve can allow us to master interplanetary travel.

If we add NPP potentials to the S-curve, we get even broader technological frontiers. First-generation NPP designs such as Project Orion (discussed in Chapter 3) can offer extremely high payloads for launch vehicles (on the order of thousands of tons per launch!) as well as specific impulses on the order of 1,000 seconds, a much higher potential than chemical rockets, which essentially eliminates any concern about payload weight and demolishes the largest barrier to current large-scale space operations—space power—today. Even a first-generation NPP launch/space operations vehicle would be a quantum leap in American space power that would forever redefine how we think of space travel. However, investigations into the theoretical maximum capabilities NPP-style propulsion can provide offers the possibility of building truly *Battlestar Galactica* or *Enterprise*-style vessels capable of reaching 10 percent of the speed of light! These would truly be starships that could travel to our closest star in 44 years, no warp drive required. Both NTR and NPP technologies are revolutionary innovations that we know how to build. If we transitioned American space power from our exhausted chemical rocket S-curve to the nuclear S-curve, we could achieve an immediate improvement in capability and begin on a path, with only incremental and evolutionary improvements necessary, that would allow us to become masters of our solar system and begin to explore our local interstellar neighborhood as well. A revolution in space power ability indeed!

But birthing this revolution will require nothing short of a revolutionary war. The nuclear technology essential to begin the nuclear space propulsion S-curve is well understood. Indeed, most of it has been worked out already, and it would take but a little effort to relearn the technology developed in the 1960s if our current engineers are not yet capable of developing the first generation of operational nuclear space vehicles. The difficulty of the space power revolution will be almost entirely political. The reason the atomic space program of the 1960s never materialized was due in part to fiscal pressures and the dulling of the space race after the Moon landings, but was mostly due to the public reaction against all things nuclear in the late 1960s and culminating in the protests against nuclear technology in the 1970s. Today, when even a relatively small and harmless radioisotope thermoelectric generator (or RTG, essentially a nuclear heat lamp) on a deep-space probe can elicit large protests from environmentalist groups, the massive reactors necessary for an NTR rocket are almost unthinkable. Exploding hundreds of nuclear bombs for an interplanetary Orion-type NPP vehicle would send many anti-nuke activists into epileptic fits. However, it is only through these activities that we will likely ever accomplish anything of note in space. Space power advocates will have to take their case to the American people and convince them that a nuclear space program is both safe and immensely valuable and that the protests of the inevitable antinuclear activists are irrational and devastating to any future in space worth having. If you want warp drive, you have to be okay with the nuclear warp core. Only by eliminating the antinuclear space bias in today's society can we have a robust space program and a mature space power.

Thus, the first phase of the American space power road map must make the jump from a chemical-powered space program to a nuclear-powered space program that is far less technologically limited. Doing so will offer us at least four immediate advantages. Firstly, the first generation of spacecraft developed using nuclear power will offer a massive increase in ability that would open up many possibilities in space virtually overnight. These opportunities would provide a large shot in the arm to generate excitement and optimism in the space program as old limitations will give way to revolutionary possibilities. Secondly, the jump to a nuclear space program will offer an intellectual breakthrough that will allow our space professionals to cease thinking small in space and begin to think big in space operations. Instead of believing the future is in microsatellites around low Earth orbit, we could think in terms of lunar and Martian colonies again. Instead of focusing inward, we would look out to the solar system for challenge and adventure. Thirdly, engineering and operational experience we get from operating these first-generation nuclear engines will begin to add to our corporate understanding of how to operate systems all along the nuclear S-curve. Just as safety rules learned through the hardearned experience of operating dangerous steam-powered vessels in the 19th century have immediate application today, so would the lessons learned today in nuclear space vehicle operations will likely be directly applicable to deep space operations hundreds of years hence, because the nuclear S-curve will likely sustain human space ventures for centuries to come.

Finally, confronting the challenges of a vastly new technology and the immense opportunities the technology opens will lay the foundation for the next two phases of our space power road map. With this new nuclear technology we will need to address both how we will use the new technology and how we will train and equip the men and women who will operate this new technology. The next phase of the road map will be explaining how we will exploit this new nuclear hammer operationally: the peaceful strategic offensive.

Operational Line of Effort: The Peaceful Strategic Offensive

> Naval strategy has indeed for its end to found, support, and increase, as well in peace as in war, the sea power of a country.—Admiral Alfred T. Mahan, U.S. Navy[25]

Seafaring nations are perpetually on the offensive in times of peace and war alike, according to Holmes and Yoshihara. Naval strategy does not start at the beginning of war and disappear with its end.[26] During war, navies can force open access to hostile markets and bases while denying access to the adversary. In peacetime, nations bolster sea power through "acquiring strategic geographic features" that will assist in expanding and defending access to the sea's riches, including efforts to reach new markets and increase available existing bases through diplomacy or construction of new ones. The development of the British Empire as a worldwide maritime empire was the result of hundreds of years of British offensive naval strategy as it opened and developed new markets for sea power in the name of national prosperity.[27]

Naval strategy differed from maritime strategy, wrote Mahan, because it had "for its end to found, support, and increase, as well in peace as in war, the sea power of a country." Finding and securing strategic geographic nodes was one way to bolster sea power in peacetime, as were efforts to hold open access to markets and bases. A nation intent on sea power was perpetually on the offensive, in wartime and peacetime alike.[28]

Space strategy is also dedicated to expanding the space power of a country in peace as well as in war. The end of space strategy is wealth from space, and it requires strategic access to space. And, as in sea power, strategic access is acquired through *strategic offensives*. Strategic offensives can be through violent or peaceful means, but dominant space powers should always be on the strategic offensive. Today, there is not much reason for kinetic strategic offensives because there is no strategic access in space that is effectively blocked by military power. Therefore, there is no denied access to be liberated through military means, nor any real reason to deny others access to space in any capacity other than in potential should a war in the future demand it. So how are we to expand access to space? Fortunately, the *peaceful strategic offensive* is not only available but is already supported by a significant group of space enthusiasts, though they think of the peaceful offensive in slightly different terms.

The space enthusiast community often speaks of space settlement, space industrialization, and creating a spacefaring society. All of these mean spreading the human ecosystem and economy to space, permanently. In other words, they are advocating a peaceful strategic offensive to rapidly increase national and international strategic access to space. Industrialization of the near Earth orbits in the form of tourist space stations, propellant depots, and manufacturing and scientific stations increases the strategic access to Earth orbit. Lunar colonies and the transportation infrastructure built to get there will rebuild a bridge to the second human world that humankind has lost for almost half a century. Mars enthusiasts who want to fashion a whole new civilization are championing probably the largest strategic access gain in any environment at any time in human history. All of these projects can be considered potential peaceful strategic offensives—they are "offensives" that intend to claim access to new territories for human commerce.

The lofty goals of the enthusiast community are really the only constituency championing the peaceful strategic offensive in space. The military, for the most part, wants to focus on military denial capabilities and is wrongly content with the current but paltry level of strategic access to space. Even investing in the very important access multiplier of low-cost launch vehicle technology is frowned upon in military circles. Expanding commercial space capabilities is little short of fantasy to most military leaders. NASA management, while talking the talk of the space enthusiast, is seen by many to prefer to spend their limited money to keep their bloated bureaucracy employed in strategically dubious enterprises. The enthusiast community, and the commercial space companies that some of the enthusiasts are starting such as SpaceX and Virgin Galactic, are the only champions of the strategic offensive now and for the immediate future.

However, how do we know whether a space development project truly advances strategic access and a space power's ability? A project advances strategic access if it will open a legitimate market, area, or natural resource to exploitation by commerce to generate wealth on a permanent and cost-effective basis. A "flags and footprints" mission to Mars or the Moon that does not blaze a trail that makes it physically and financially easier for private commercial missions to follow (say, with leaving permanent space refueling platforms or some other transportation infrastructure) does not increase strategic access. Strategic access is not like a flag carrier storming an enemy trench in World War I. Simply placing a marker somewhere and saying "we've been there" does not advance strategic access. Instead, any mission must blaze a path that others can follow to generate wealth. The scout must not simply go, but offer help to the prospectors and settlers who follow.

Space projects must be screened for their potential to expand strategic access by answering the following questions: (1) Will it open a new area, market, or natural resource for commerce and economical exploitation? and (2) Will it be cost effective, financially sustainable, and promise profitable returns either for itself or budgeted (not simply vaguely "planned") follow-on missions? If the answer to both of these questions is yes. then the project in question may be valuable. Profitable returns can come in material wealth, political power, or scientific knowledge. The most valuable is material wealth. Lucre is the lifeblood of projects and nations. Political power exists but is often transitory, such as the goodwill generated by the Apollo program. Science is also often overvalued by space scientists and enthusiasts. Scientific knowledge is only so valuable in a space power sense in that it has a serious potential to provide expanded markets, resources, or transportation possibilities. In short, is the knowledge amenable to applied research and development? If the knowledge is not, then it cannot be properly understood as space wealth in a strategic sense, and is nothing more than political patronage to the scientist or science enthusiast community.

Any peaceful strategic offensive must attempt to increase strategic access. Therefore, not all proposed missions can be considered strategic offensives.

Regardless of how strong a constituency is lobbying for a particular mission or operation, if the project does not open new areas for commercial exploitation in a cost-effective and financially stable manner it cannot be considered a strategic offensive and does not advance space power in the Mahanian sense. If the project does answer these essential questions affirmatively, then we can be assured that proceeding on the project will increase humanity's and the nation's strategic access to space—thus increasing the wealth, power, and well-being of the nation and establishing a better foothold for the human race's future in the heavens.

A potential objection to using Admiral Mahan's sea power theories as a foundation for space power theory is the perceived violence committed in the late 19th and early 20th centuries due in part to Mahan's writings. Wars such as the Spanish-American War of 1898, the expensive Dreadnought-era naval buildup between Britain and Germany before the Great War, and the American colonial takeover of Puerto Rico, Hawaii, and the Philippines can honestly be partially attributed to Mahan's vision of sea power. However, as we have seen, Mahan's theory does not champion aggressive war, but peaceful commerce. Much of the navalist aggression that occurred in Mahan's time was perpetrated by people other than Mahan and used Mahan only as a justifiable pretext for actions they would have accomplished anyway. Also, a significant advantage of a Spacer navalism is that, while sea power in the 19th century was essentially a zero-sum game where the globe was settled and resources were by necessity gained from territorial conquest by diplomacy or arms, space power for the foreseeable future is an open system, with wealth, space, and freedom available to all bold enough to step out and reach it. Perhaps the best thing about space power is that a national colonial impulse will now animate dead worlds rather than bring the enslavement of the weak by the strong as it has in the past, taking what many consider to be an evil impulse and turning it to the service of life. There is something noble and redemptive about this change and the awesome potential of the peaceful strategic offensive. Therefore, it is imperative for American space power to prosecute a peaceful strategic offensive in space. To accomplish the offensive, however, will require men capable of reaching for the stars ... and grabbing them!

Organizational Line of Effort: Developing the Spacer Officer

The scenarios presented above, and the investigation of the usefulness of some potential technological development paths for each scenario, lead us to consider the peaceful strategic offensive—rapid expansion of space access through developing better space transportation—as the best potential route for the Spacers to take. With a strategy in hand, now we must turn to developing the people who can implement that strategy and form the new space service.

The General Theory of Space Power views development primarily in terms of idea generation. Idea generation involves both combinations of elements that provide access (grammar), and transformers that channel general space power into applied national power (strategy). However, idea generation only provides the raw material for development. In order for development to be successful, the ideas generated must improve useful access and advance an agent's space power ability. This implies that development requires both idea generation and strategic analysis of ideas generated.

Idea generation and element improvement will generally be accomplished by scientists and engineers, though anyone can contribute. Skill in strategic analysis, however, may be the most important contribution space military officers—the Spacers—may be able to provide to the effort to prepare for the Spacers' War. Drs. Stefan Possony and Jerry Pournelle, and Air Force Colonel Frank Kane, devote a great deal of time to exploring the concept of the strategic analyst in their book *The Strategy of Technology*. The authors explore their concept of Technological War, of which the Spacers' drive to develop space power will be a major theater, which they describe as:

> Technological warfare is the direct and purposeful application of the national technological base and of specific advances generated by that base to attain strategic and tactical objectives. It is employed in concert with other forms of national power. The aims of this kind of warfare, as of all forms of warfare, are to enforce the national will on enemy powers; to cause them to modify their goals, strategies, tactics, and operations; to attain a position of security or dominance which assists or supports other forms of conflict techniques; to promote and capitalize on advances in technology to reach superior military power; to prevent open warfare; and to allow the arts of peace to flourish in order to satisfy the constructive objectives of society....[29] International technological competition can sometimes reach levels best described as economic warfare, and the outcomes of these competitive struggles can have surprisingly long range effects on the decisive military Technological War.[30]

Since Technological War is a different way of viewing international relations and classical warfare, the concepts of victory and defeat are somewhat different. They continue:

> Victory in the Technological War is achieved when a finite game participant (i.e., one who wishes to bring the game to an end by winning it) has a technological lead so far advanced that his opponent cannot overcome it until after the leader has converted his technology into decisive weapons systems. The loser may know

that he has lost, and know it for quite a long time, yet be unable to do anything about it....

In summary, proper conduct of the Technological War requires that strategy drive technology most forcefully; that there be an overall strategy of the Technological War, allocating resources according to well-defined objectives and an operational plan, not merely strategic elements which make operational use of the products of technology. Instead of the supply officer and the munitions designer controlling the conduct of this decisive war, command must be placed in the hands of those who understand the Technological War; and this requires that they first understand the nature of war.

Lest the reader be confused, we do not advocate that the Technological War be given over to the control of the scientists, or that scientists should somehow create a strategy of technological development. We mean that an understanding of the art of war is more important than familiarity with one or another of the specialties of technology. It is a rare scientist who makes a good strategist; and the generals of the Technological War need not be scientists any more than the generals of the past needed to be good riflemen or railroad engineers.

Like all wars, the Technological War must be conducted by a *commander* who operates with a strategy. It is precisely the lack of such a strategy that brought the United States to the 1970's low point in prestige and power, with her ships seized across the world, her Strategic Offensive Forces (SOF) threatened by the growing Soviet SOF—and with the United States perplexed by as simple a question as whether to attempt to defend her people from enemy thermonuclear bombs, and unable to win a lesser war in South East Asia.

We had neither generals nor strategy, and muddled through the most decisive conflict in our national history.... There always were exceptions to this unsatisfactory record of American performance. General Bernard Schriever created a military organization for strategic analysis which was responsible for our early commanding lead over the Soviets in ballistic missiles, despite the fact that the U.S. had allowed the U.S.S.R. many years' head start in missile development after World War II. The Air Force's Project Forecast and later Project 75, was an attempt to let strategy react to, then drive, technology; these, too, were creations of General Schriever's.[31]

While the Spacers' War may become hot, the buildup to it will probably be the hot peace characterized by Technological War. Thus, Spacers will need to be both masters of space in the Technological War realm as well as the hot realm. This will require deeply understanding the attributes of the Technological War.

ATTRIBUTES OF TECHNOLOGICAL WAR

Up to the present moment, technological warfare has largely been confined to pre-hot war conflict. It has been a silent and apparently peaceful war, and engagement in the Technological War is generally compatible with the strong desires of most of our people for "peace." The temporary winner of the Technological War can, if he chooses, preserve peace and order, act as a stabilizer of international affairs, and prevent shooting wars—continue the Technological War as an infinite game.

There could be a different outcome. If the side possessing a decisive advantage sees the game as finite, the victor can choose to end the game on his own terms. The loser has no choice but to accept the conditions of the victor, or to engage in a shooting war which he has already lost.

Technological War can be carried on simultaneously with any other forms of military conflict, diplomatic maneuvers, peace offensives, trade agreements, detente, and debacle. It is the source of the advanced weapons and equipment for use in all forms of warfare. It renders cold war activities credible and effective. Technological warfare combined with psychosocial operations can lead to a position of strategic dominance.[32]

Strategic (logic) dominance in space must be the ultimate goal of the Spacers in the lead up to the Spacers' War. Indeed, sufficiently dominant space forces may ultimately prevent the Spacers' War from happening at all. As Possony et al. continue, they stress the importance of building an organization that can successfully conduct technological warfare:

> Our misunderstanding of the Technological War is illustrated by our failure to build an organization for conducting technological warfare. The review of the annual budget and of individual projects in basic research, in applied research, in development, and in procurement is the only process by which our technological development is controlled directly. Other influences such as the statements of requirements and the evaluation of military worth are felt only at the level of individual projects. Overall evaluation of the research and development effort and of its relations to strategy is rudimentary.
>
> An example of how irrelevant factors influence our efforts, and perhaps one of the decisive signs of the times: the January 20, 1969 issue of *Aviation Week and Space Technology*, the most influential journal in the aerospace field, included a report entitled "Viet Lull Advances New Weapons." The article makes clear that the budgetary funding level of many advanced new weapons systems, including research and development, basic technology, and actual system procurement, is largely dependent on the continuation of a "lull" in the Vietnam war. Given a proper strategy for the Technological War and proper command of our efforts, the title should read "Advanced New Weapons End Vietnam War."[33]

Even though they did not prevent the Pacific War, the navalists do offer a fine model with which to organize for Technological War. Adopting a similar structure to the Navalist NWC–GB–OpNav Triangle devoted to Path 5 development is well suited for conducting Technological Warfare. Developing a synergistic triangle of a Space War College, a Space General Board, and strongly linking them both to the Space Operations Command in the United States military space forces is both a historically proven and theoretically effective organizational approach for prosecuting Technological Warfare in space.

A SPACE WAR COLLEGE

Establishing a Space War College dedicated to higher-level research on space power issues, education of senior Spacer leaders, and war gaming advanced space power ideas is a nearly essential first step in establishing the Spacer culture and putting the United States on a path to developing a mature space power. The Naval War College was the first true war college in the United States, "professionalized" the naval officer vocation, and became the hub of navalist thinking. The officers of the Air Corps Tactical School (now the Air University) did not invent the theories such as strategic bombing that brought air power into maturity, but they did refine previous thinking into useful doctrines that enabled heavy bombers to be built, successful air campaigns to contribute to war efforts, and enable the United States Air Force to become an independent service. Likewise, the Space War College must serve to study, test, war game, and develop today's often divergent thinking on space power and coalesce it into usable tactics, strategies, matériel technical requirements, and a coherent space theory of victory. Most of all, it must serve to educate and motivate today's space officer to become tomorrow's new navalist, the Spacer officer. Possony et al. explains the purpose of a war college in terms of Technological War:

> As General Beaufre has pointed out, the strategist must not limit himself to what is possible; he must find ways to do what is necessary. Wishing for a technological capability will not necessarily give us one, but the history of technological development, particularly of weapons, leads us to believe that identification of a technological requirement increases the likelihood of fulfilling it....
>
> What we must do is encourage strategic thought, particularly among younger officers, and ensure promotion for officers who show genuine strategic talents. This nation has always been fearful of a general staff, falsely identifying this useful military instrument with Prussia and Nazi Germany and supposing it to be incompatible with democratic institutions. When the structure of a general staff corps is explained, not one American in a thousand recognizes what it is; yet he no longer fears it when he does understand it. There may be good reasons for rejecting the general staff concept, but we venture to suggest that it be rejected for something better than a pipe dream such as that which was brought to an end by the historic event at Kitty Hawk.
>
> In fact, the general staff corps concept is this: at an early stage in their careers, certain young officers are selected as potential strategists, intelligence experts, and staff officers. Management of their careers is then given to the general staff; they are posted to staff assignments and schools where they study war, strategy, tactics, military doctrine, and history. School assignments are alternated with service in the field and with such special arms as artillery, infantry, and armor. They remain in the general staff corps until they are thought to be unsuitable for it, whereupon they can either be transferred to one of the line services or retired. During their

careers in the corps, the selected officers alternate between appointments to general staff headquarters and its specialized branches—such as logistics, and attaché duties—and appointments in the field, where they serve as chiefs of staff to the field commanders of successively larger units. Thus, commanders learn to command and staff officers learn the functions of staff work. Commanders and staff officers each have their own paths of promotion, and are not in competition with each other until they come to the highest positions. Even there, competition may be kept to a minimum because staff officers often make good commanders above the corps level.

This, in brief, is the general staff corps system. It produces officers who have considerable knowledge of strategy; it requires them to be familiar with the operations of the military services and the tactics of the field forces; and it encourages them to think in intellectual rather than command terms. The system has been proved to be effective, although it is subject to improvements.

Whether it be through the general staff concept or some other, we must find ways of selecting, training, promoting, and rewarding strategic talent and placing it in positions where it would be able to formulate successful strategy. Without strategists we will have no strategy. Yet it is strategy that is our greatest need in the Technological War.[34]

The war college has been the traditional home for modern strategic development in the United States military. Space strategist Navy Commander John Klein identifies that the Space War College is both critical and a great first step for many reasons:

Although a separate space service is not currently needed, a Space War College should be established. Only by changing the mindset of the professional warfighter can space be acknowledged as a distinct and co-equal medium of warfare—like land, naval, and air warfare. To do this, a separate war college should be established. This would indicate that space warfare is also a category of warfare in which vital interests must be defended and protected. The establishment of a Space War College is the only method, short of establishing a separate space service, by which warfighters and policy makers will recognize that more thought and effort should be expended to protect our nation's interests and security in space. Granted that there will be startup and operational costs associated with the establishment of such a school, such a cost is quite small in comparison with how much space-reliant commerce and trade currently takes place domestically and abroad.[35]

Thus, the Space War College both plays an essential part in strategy formation and is a relatively inexpensive way to begin laying the foundation for the essential Spacer culture. The Space War College also need not be developed completely from scratch. Leveraging current military schools such as the Air University's School for Advanced Air and Space Studies as well as the National Security Space Institute and the Advanced Space Operations School (collectively known as the Space Education and Training Center) in Colorado Springs, Colorado, would make developing the Space War College much cheaper, and likewise even more attractive. But the Space War College is only one leg of the Spacer Triad.

THE SPACE GENERAL BOARD

The General Board concept, also discussed in Chapter 4, is not unique to the Navy. In fact, the Air Force had its own General Board in the form of the Air Service (later Air Corps) Board in the 1920s and '30s. In both cases, the General Board was a vital link that connected the theoretical work of the War College to the practical needs of the operational force. As Possony et al. describe:

> We must have at all times the in-being force necessary to win wars. This means being ready for operations at every moment in the foreseeable future while providing simultaneously the foundations for major advances in future capabilities. These are requirements that compete for resources. Our in-being capability is not static; we cannot allow it to dwindle or become obsolete. Thus, modernization of our forces must be continuous but it cannot detract from having sufficient power at any given time.[36]

Maintaining the appropriate balance between modernization and current operations will be the role of the Space General Board. The General Board would give due consideration to the ideas generated and war gamed by the War College and recommending adoption of the idea by operational space forces when appropriate. Likewise, the General Board would consider questions posed from the operational forces and direct the War College to study them as necessary. The board, like the Navy and Air Boards before, would be composed of senior and mid-level officers from both the War College and the Operational Command.

In both the Navy and Air Force cases, the early boards were generally located in close proximity to the war colleges—with mid-grade officers especially often assigned duties to both. Directing the same philosophy of staffing to the Space General Board would be easy since both the NSSI/ASOpS and Headquarters, Air Force Space Command are colocated at Peterson Air Force Base in Colorado Springs. Defining a working relationship between the two in the form of a Space General Board would be very easy organizationally and trivial logistically. With two of the Triad's legs set, we can finally complete the Spacer Triad.

THE SPACE ACADEMY

The third leg of the Spacer Triad is an Operational Command sufficiently linked to the Space War College and the General Board. Then why do we need to discuss a Space Academy? The answer is because we already have an operational space organization similar in nature to the navalists' OpNav–Air Force Space Command. AFSPC as an organization needs no real change in order to fulfill the Spacers' need for an operational command. What AFSPC needs to do in order to act as the third leg of the Spacer Triad is to commit itself to engage in mass experiments akin to the Navy's interwar Fleet Problems in order to incorporate the innovations from Path 5 space power development from the War College and General Board. Albert Nofi explains the role of the Naval Academy to the success of the Fleet Problems:

> [Speaking of the Navy's interwar Fleet Problems and the causes of their successes] Virtually all naval officers were [Naval] Academy graduates, and the fleet only had one "community," the big gun navy, with everyone—even aviators and submariners—sharing a common understanding of operational and tactical procedures, matériel, and "culture."[37]

In short, the Naval Academy engendered virtually the entire officer corps into the common navalist culture that allowed naval innovation to take place. The Naval Academy made everyone navalists. In the same way, the Space Academy is an essential tool in developing a clear space "community"—the Spacer community—that shares a common understanding of procedures, a common culture, and a common grand vision (though it's critical to encourage innovation and stymie groupthink at smaller levels). Thus, the Space Academy will be the critical node that changes AFSPC over time to become a Spacer organization. Dr. Travis Taylor recognizes the need for a central hub of space learning, and an associated culture, in his book *A New American Space Program*:

> What we need is a full up academy. We need a ... Space Academy. A centralized university type environment (I'd prefer it be in the Rocket City but I'm biased) where young people can go and get degrees in Space-Oriented Science, Engineering, and even Management. The curriculum should contain every aspect from astrophysics to rocketry to pilot training to mission control to medical training and everything else in between. It is at the United States of America Space Fleet Academy where anyone who can get accepted (through standard collegiate requirements and pilot physical status) can go to this university and get his or her space education. The degrees would all be tailored to land space jobs for graduated students. Pilot trainees could springboard into military, civil, or commercial aviation for further training and maintaining currency in pilot status. The other students (non-pilots) would acquire Bachelor of Science, Master of Science, and Doctor of Philosophy degrees in their particular fields.
>
> The output of the academy would be a generation of young people fully trained and full of enthusiasm to push forward into space and to bring on a new era of humanity's involvement in space. The more students we educate on the benefits of space exploration and utilization, the more mainstream it will become in our culture, society, and business.
>
> Private companies and university activities really cannot generate the pull them-

selves to create real space exploration programs, with real national, and even global, attention. The USA Space Fleet Academy would. It would be easy for such an organization to be set up with modest amounts of the national space exploration budget. If the budget matched what it should, compared to that of the Apollo era, it would be *very* easy. Five percent of the NASA budget could easily set up such an academy, with some to spare. Since the other services and industry would be involved, they could also become partners in the expense. Five percent of the space exploration budget currently is somewhere between $500 million and $1 billion a year. This could set up flight training capabilities, space mission simulators, laboratories, classrooms, world-class instructors, and even the more mundane logistical pieces of such an academy such as buying a campus, setting up buildings, and marketing to students.

There are pieces of the space program, like such an academy, that we have simply overlooked. If we train a Space Fleet, then soon enough that fleet will want to go into space. And they will get frustrated with the politicians if it doesn't happen. We will also train a large group of individuals on how important it is for humanity to stretch out into the heavens. That group of people will then communicate with their families and friends and spread the word. The concept of humanity in space would become as viral as being "Army Strong," or pulling for the New York Giants to win a Super Bowl.[38]

Taylor instinctively hits at the important psychological benefit the Space Academy will offer the Spacer community. Not only will young Spacer officers be developed by the Space Academy, the institution will also provide a healthy and growing political constituency for a powerful and innovative American space power program. It is for this psychological and cultural boost, as much as for high-quality young recruits trained to dominate space, that the Space Academy is an essential Spacer project.

Interestingly, Taylor concludes that this Space Academy should not train graduates specifically for military service. Instead, Taylor would have graduates sent throughout the American space industry into the military, civil, and commercial programs. One can assume that this nonmilitary focus is because Taylor is not a military man himself, but his sentiment is actually quite common. Army Lieutenant General Daniel Graham (a lead supporter of the Delta Clipper DC-X project and space-based missile defense advocate in the early 1990s) advises the same concept:

> What is needed is a Space Academy along the lines of the U.S. Merchant Marine Academy where young Americans, and perhaps youngsters from other lands, could be steeped in the disciplines pertinent to space. They could be graduated with a bachelor's degree in space science, ready for employment in aerospace industry, commercial space enterprises, or government civil and military space programs.
>
> Over the past thirty years, many space-oriented disciplines have developed: space medicine, space law, space diplomacy, space construction, and space literature, to name a few. There would be no difficulty whatever in establishing a space-oriented

curriculum for the cadets of a Space Academy meeting all necessary criteria for granting of degrees.

Further, the standards for entry into such an academy could be set very high and still not accommodate all young people who would apply and qualify. A student body with computer literacy, technical aptitude, physical fitness and high personal standards would be easily assembled, as well as a competent, highly motivated faculty.

If the general space policies I recommend were pursued, graduates of such a Space Academy would be in great demand, and the aspirations of young people interested in a career in space-related activities could be channeled effectively.[39]

Spacers need to be everywhere in the United States space community. Therefore, Space Academy graduates will need to be everywhere. Space power has economic, political, and military dimensions and healthy development will embrace innovation and require effective management over all three. It makes complete sense for the Space Academy to allow graduates to pursue careers in any dimension of space power they choose.

The U.S. Merchant Marine Academy model offers a unique advantage for the Space Academy to consider. USMMA midshipmen are given a full scholarship just as U.S. military academies are free to students. America receives her benefits from training USMMA graduates by requiring that graduates spend a set number of years in an approved seafaring industry of their choice (commissioning into an active duty military service is a choice as well). However, all graduates of the USMMA who do not go on to active duty military status are commissioned into the U.S. Naval Reserve as officers. Therefore, all USMMA graduates are military officers who can be called upon in times of national emergency. Likewise, the U.S. Space Academy should allow its graduates to go wherever they are most interested in the U.S. space industry, while ensuring that every USSA graduate has a Spacer uniform in their closet and is ready to answer the nation's call if the necessity ever comes.

With a Space War College, Space General Board, and Space Academy working together to form an effective Spacer Triad for Path 5 space power development, the organizational line of effort for our space power plan will conclude. Following these three lines of effort: the technical (nuclear space propulsion), the operational (the peaceful strategic space offensive), and the organizational (the Spacer Triad), the United States will be on a solid, aggressive, and rewarding path to true space power development.

Conclusion

By understanding both the Logic and Grammar of Space Power and applying their lessons, an innovative and nationally valuable era of American space power will be unleashed. Today, this type of American space power does not exist. As General Graham lamented:

> In May 1967, Krafft A. Ehricke, a prominent space scientist and pioneer, told the American Astronautical Society: "Utility, that is, the capability of satisfying human needs, is the only foundation strong enough to sustain an ever growing superstructure of explorative astronautics in the decades and centuries to come."
> Ehricke was absolutely right. But U.S. government space programs did little to follow his advice. Instead, space endeavors became more and more a government monopoly, with little regard for utilitarian applications—"satisfying human needs."[40]

The ultimate utility of the space power model this book has developed and explored is to ensure that human needs must be the end to all space effort. These ends can be economic, political, or military in nature, but all activity must satisfy human need.

Dr. Lee Valentine, director of Princeton's Space Studies Institute, summed up his space road map with one statement: Mine the Sky, Defend the Earth, Settle the Universe.[41] Regardless of what one thinks about its merits, it is an elegant summary of a complete space power platform because, consciously or not, Dr. Valentine posits an economic platform, a military platform, and a political platform for space power—hitting all of the points of the Logic of Space Power. Should the United States care to adopt Valentine's prescription for space power or not, by adhering to the Logic and Grammar of Space Power, American's position as the world's dominant space power will be assured.

Chapter Notes

Introduction

1. *The Space Report* 2013. The Space Foundation. Colorado Springs, Colorado. 2013, p. 74.

2. Ibid., p. 71.

3. Ibid., p. 69.

4. John J. Tkacik, "China Space Program Shoots for Moon," *The Washington Times*, 8 January 2010. http://www.washingtontimes.com/news/2010/jan/08/china-eyes-high-ground/?page=2#ixzz2czkK0BhW (accessed 2 September 2013).

5. Joan Johnson-Freese, "Will China Overtake America in Space?" CNN.com. 20 June 2012. http://www.cnn.com/2012/06/20/opinion/freese-china-space.

6. Robert Bigelow, speech given in Las Cruces, NM, 19 October 2011. http://bigelowaerospace.com/The_New_China_Syndrome.pdf. (accessed 20 February 2014).

7. Erik Seedhouse, *The New Space Race: China vs. the United States*. Chichester, UK: Praxis, 2010, p. 220.

8. Futron Corporation. *Futron's 2012 Space Competitiveness Index*. Bethesda, MD: Futron, 2012, p. 2. The 2012 report is used because the 2013 SCI report is proprietary information and distribution is limited by Futron Corporation.

9. Ibid., p. 10.

10. Ibid., p. 50.

11. Ibid., p. 122.

12. Joseph Schumpeter, *The Theory of Economic Development*. Cambridge, MA: Harvard University Press, 1934. Reprint, New Brunswick, NJ: Transaction Press, 2008, p. 64 (note).

13. J.C. Wylie, *Military Strategy: A General Theory of Power Control*. Annapolis, MD: Naval Institute Press, 1967, p. 31.

14. Harold Winton, "On the Nature of Military Theory." In Lutes and Hayes, eds. *Toward a Theory of Spacepower*. Washington, DC: National Defense University Press, 2011, p. 22.

15. Winton, p. 23.

16. Winton, p. 32.

Chapter 1

1. Colin S. Gray, "The American Way of War." In McIvor, ed., *Rethinking the Principles of War*. Annapolis, MD: Naval Institute Press, 2005, p. 21.

2. William Mitchell, "Winged Defense." In Jablonsky, ed. *Roots of Strategy, Book Four*. Mechanicsburg, PA: Stackpole Books, 1999, p. 425.

3. Frederick Baier, *50 Questions Every Airman Can Answer*. Maxwell AFB, AL: Air University Press, 1999, p. 7.

4. Baier, p. 7.

5. Philip S. Meilinger, *10 Propositions Regarding Air Power*. Washington, DC: Air Force History and Museums Program, 1995, p. 20.

6. David Lupton, *On Space Warfare: A Space Power Doctrine*. Maxwell AFB, AL: Air University Press, 1988, p. 7.

7. Mitchell, p. 425.

8. Lupton, p. 7.

9. Carl von Clausewitz, *On War*. Howards and Paret (editors). New York: Knopf, 1999, p. 731.

10. James Holmes and Toshi Yoshihara, "Mahan's Lingering Ghost." *U.S. Naval Institute Proceedings*, December 2009. Annapolis, MD: Naval Institute Press, 2009, p. 41.

11. Alfred T. Mahan, *The Influence of Sea Power Upon History, 1600–1783*. Cambridge, MA: John Wilson and Sons, 1890, p. 28.

12. *Influence*, p. 44.
13. James E. Oberg, *Space Power Theory*. Colorado Springs, CO: United States Space Command, 1999, p. 44.
14. Ibid.
15. *Influence*, pp. 28–29.
16. Holmes and Yoshihara, p. 43.
17. Ibid., p. 41.
18. Alfred T. Mahan, *Retrospect and Prospect*. Boston, MA: Little, Brown and Co., 1902, p. 246.
19. Holmes and Yoshihara, p. 43.
20. Ibid., 44.
21. Ibid., 43.
22. Ibid.
23. Susan Ward, "Business Plan." http://sbinfocanada.about.com/cs/startup/g/businessplan.htm (accessed 20 February 2014).
24. Joan Johnson-Freese, *Space as a Strategic Asset*. New York: Columbia University Press, 2007, p. vii.
25. Joseph S. Nye, *Soft Power*. New York: Public Affairs, 2004, p. 5.
26. I.B. Holley, *Technology and Military Doctrine*. Maxwell AFB, AL: Air University Press, 2004, p. 3.
27. Ibid., p. 2.
28. Joseph Schumpeter, *The Theory of Economic Development*. Cambridge, MA: Harvard University Press, 1934. Reprint, New Brunswick, NJ: Transaction Press, 2008, p. xix. (Introduction by John E. Elliott.)
29. Ibid., p. 66.
30. Ibid.
31. Everett C. Dolman, *Pure Strategy*. London: Frank Cass, 2005, p. 42.
32. Ibid., p. 42. Emphasis original.
33. John S. Lewis, *Mining the Sky*. Reading, MA: Addison-Wesley, 1996, p. ix. Emphasis original.
34. Ibid., p. ix–x.
35. Ibid., p. x.
36. Stephan T. Possony and J.E. Pournelle, *The Strategy of Technology*. Cambridge, MA: Dunellen, 1970, pp. 45–46.
37. William H. Goetzman, *Exploration and Empire*. New York: History Book Club, 1966, pp. 600–601.
38. Holmes and Yoshihara, p. 43.

Chapter 2

1. Schumpeter, p. 74.
2. Ibid., p. 75.
3. Ibid.
4. I.B. Holley, p. 88.
5. Ibid., p. 105.
6. Ibid. Italics original.
7. Richard V. Adkisson, "The Original Institutionalist Perspective on Economy and Its Place in a Pluralistic Paradigm." *International Journal of Pluralism and Economics Education.* Vol 1, No. 4 (2010), p. 361.
8. Ibid., p. 362.
9. Quoted in Holley, p. 94.
10. Quoted in Holley, p. 105–106. Emphasis added.
11. Wernher Von Braun (White, Henry, trans.). *The Mars Project*. Urbana: University of Illinois Press, 1953.
12. W. Patrick McCray, *The Visioneers*. Princeton: Princeton University Press, 2013, p. 3.
13. Mark Erickson, *Into the Unknown Together*. Maxwell AFB, AL: Air University Press, 2005, p. 61.
14. Ibid.
15. Ibid., p. 61.
16. Ibid., p. 60.
17. Ibid., p. 61.
18. Ibid., p. 156.
19. *Project Horizon: A U.S. Army Study for the Establishment of a Lunar Military Outpost*. Volume I. U.S. Army Ballistics Missile Agency. 8 June 1959, p. 2. http://www.history.army.mil/faq/horizon/Horizon_V1.pdf (accessed 20 February 2014).
20. James R. Ronda, *Beyond Lewis & Clark: The Army Explores the West*. Tacoma: Washington State Historical Society, 2003, p. 1.
21. Ibid., p. 28.
22. Ibid., p. 94.
23. Nathaniel Philbrick, *Sea of Glory: America's Voyage of Discovery, The U.S. Exploring Expedition, 1838–1842*. New York: Viking, 2003, p. xix.
24. Simon P. Worden and John E. Shaw, *Whither Space Power? Forging a Strategy for the New Century*. Maxwell AFB, AL: Air University Press, 2002, p. 110.
25. Ibid., p. 114.
26. Ibid., p. 116.
27. Quoted in Erickson, pp. 77–78.
28. Erickson, p. 105.
29. Quoted in Erickson, p. 107.
30. Quoted in Robert Godwin, editor, *Dyna-Soar: Hypersonic Strategic Weapons System*. Burlington, ON: Apogee Press, 2003, pp. 203–204. Emphasis added.
31. Erickson, p. 108.
32. Ibid., p. 109.
33. Quoted in Erickson, p. 110. Emphasis added.
34. To paraphrase physicist and astronomer

Dr. Michael Papagiannis somewhat out of context. He originally used this logic to defend study of UFOs.
35. Erickson, p. 110.
36. Ibid.
37. Quoted in Erickson, p. 109.
38. Quoted in Erickson, p. 106.
39. Quoted in Erickson, pp. 106–107.
40. George Dyson, *Project Orion: The True Story of the Atomic Spaceship*. New York: Owl Books, 2002, p. 191.
41. George Dyson, "Project Orion: Deep Space Force." 4 March 2008. http://makezine.com/magazine/make-13/project-orion-deep-space-force/ (accessed 20 February 2014).
42. "Deep Space Force," p. 182.
43. Quoted in "Deep Space Force," pp. 182–183. Emphasis added.
44. *Project Orion*, p. 206.
45. "Deep Space Force," p. 183.
46. Ibid.
47. *Project Orion*, p. 284.
48. Ibid., pp. 285–286.
49. Quoted in Erickson, p. 67.
50. John Tirpak, "The Space Commission Reports." *AIR FORCE Magazine*. March 2001, p. 34.
51. Ibid.
52. J.C. Wylie, *Military Strategy: A General Theory of Power Control*. Annapolis, MD: Naval Institute Press, 1967, p. 150.
53. Holley, pp. 115–116.
54. Holley, pp. 113, 116–117.
55. Adkisson, p. 367.
56. Holley, pp. 117–118.

Chapter 3

1. Stephan T. Possony and J.E. Pournelle, *The Strategy of Technology*. Cambridge, MA: Dunellen, 1970, pp. 45–51.
2. Ibid., p. 48.
3. Ibid., p. 45.
4. Ibid., p. 46.
5. Ibid., pp. 46–47.
6. Ibid., p. 46.
7. Schumpeter, p. 223.
8. Ibid., pp. 13–15.
9. John M. Collins, *Military Strategy*. Washington, DC: Potomac Books, 2008, p. 224.
10. Possony, Pournelle, and Kane, pp. 4–5.
11. Ibid., p. 16.
12. Ibid., p. 10.
13. Stefan Possony, Jerry Pournelle, and Francis Kane. *The Strategy of Technology*. 1997 Revision (electronic). Chapter 1. http://www.

jerrypournelle.com/slowchange/Strat.html (accessed 22 September 2013).
14. Richard N. Foster, *Innovation: The Attacker's Advantage*. New York: Summit Books, 1986, pp. 31–32.
15. Possony and Pournelle, p. 50.
16. Michael I. Handel, *Masters of War*. Third Revised Edition. New York: Routledge, 2001, p. 165.
17. Clausewitz, pp. 625–626.
18. Ibid., p. 528.
19. Handel, p. 187 (minor changes to align with author's Figure 2, emphasis original).
20. Holley, p. 14.
21. Possony and Pournelle, p. 12.
22. Handel, p. 187.
23. Clausewitz, p. 80.
24. Ibid., p. 570 (emphasis original).
25. Ibid., pp. 572–573 (emphasis original).
26. This section is adapted from Foster, pp. 265–277.
27. Foster, p. 267.
28. Possony and Pournelle, p. 55.
29. Holley, pp. 5–6.
30. Ibid., p. 178.
31. James Rickards, *Currency Wars: The Making of the Next Global Crisis*. New York: Penguin, 2011, p. 151.
32. Ibid., p. 149.
33. Ian Fletcher, *Free Trade Doesn't Work*. Sheffield, MA: Coalition for a Prosperous America, 2011. p. 15.
34. Robert W. Shufeldt, *The Relation of the Navy to the Commerce of the United States*. Washington, DC: John L. Gink, 1878, p. 5.
35. Ibid. Emphasis original.
36. Ibid., p. 6. Emphasis added.
37. Ibid., pp. 6–7.
38. Ibid., p. 7.
39. Ibid., p. 8.
40. Ibid.
41. Ibid., pp. 3–4.
42. Ibid., p. 4.
43. Ralph Gomory and William Baumol, *Global Trade and Conflicting National Interests*. Cambridge, MA: MIT Press, 2000, p. 16.
44. Ibid., p. 4.
45. Ibid., p. 5.
46. Ibid., p. 69.
47. Ibid., p. 70.
48. Michael Porter, *The Competitive Advantage of Nations*. New York: Free Press, 1990, p. 119.
49. Fletcher, p. 241.
50. Ibid., p. 233. Emphasis original.
51. Wingo Dennis, "Economic Development of the Solar System: The Heart of a 21st-

Century Spacepower Theory." In Lutes, et al., eds., *Toward a Theory of Spacepower: Selected Essays*. Electronic Version. http://www.ndu.edu/press/lib/pdf/spacepower/spacepower.pdf (accessed 22 February 2014), p. 174.

52. James Dewar, *The Nuclear Rocket: Making the Planet Green, Peaceful, and Prosperous*. Burlington, ON: Apogee Books, 2009, p. 37.

53. Ibid.

54. Ibid.

55. Ibid.

56. Ibid.

57. Dewar, p. 37.

58. Ibid., p. 38.

59. Ibid., p. 39.

60. Ibid., p., 38.

61. Ibid.

62. Ibid., p. 37.

63. Michael Okuda and Denise Okuda. *The Star Trek Encyclopedia*. Revised Edition. New York: Pocket Books, 1999, pp. 204–205.

64. K.F. Long, *Deep Space Propulsion*. New York: Springer, 2012, p. 191.

65. Jerry Sellers, *Understanding Space: An Introduction to Astronautics*. Revised Second Edition. Boston: McGraw-Hill, 2004, p. 361.

66. Dewar, p. 53.

67. Ibid.

68. Ibid.

69. Stann Gunn, John Napier, and James Dewar, "Development of First-through-Fourth-Generation Engines," reprinted in Dewar, *The Nuclear Rocket*, p. 185.

70. Dewar, p. 47.

71. Paraphrased by Dewar, p. 53. Note the deletion of Dewar's original fission fragment (Orion-class) engine, which explodes small nuclear fission bombs behind the ship to provide propulsion.

72. Dewar, p. 53.

73. Ibid.

74. Frank O. Braynard, *S.S. Savannah: The Elegant Steamship*. Atlanta: University of Georgia Press, 1963. Reprint, Garden City, NY: Dover, 1988, p. xi.

75. Ibid., p. 211.

76. Ibid., p. 213.

77. Ibid.

78. Edward Radlauer and Ruth Radlauer. *Atoms Afloat! The Nuclear Ship* Savannah. London: Abelard-Schuman, 1963, p. 17.

79. Radlauer and Radlauer, p. 110.

80. Dewar, p. 39.

81. Schumpeter, p. 65, emphasis added.

82. William J. Holland, Jr., "Strategy and Submarine." *United States Naval Institute Proceedings*. December 2013, pp. 48–49.

Chapter 4

1. John B.Hattendorf, et al. *Sailors and Scholars: A Centennial History of the United States Naval War College*. Newport, RI: Naval War College Press, 1984, p. 1.

2. Mark R. Shulman, *Navalism and the Emergence of American Sea Power, 1882–1893*. Annapolis, MD: Naval Institute Press, 1995, p. 151.

3. John T. Kuehn, *Agents of Innovation: The General Board and the Design of the Fleet That Defeated the Japanese Navy*. Annapolis, MD: Naval Institute Press, 2008, pp. 12–13.

4. J.A.S. Grenville, *Diplomacy and War Plans in the United States, 1890–1917*. Transactions of the Royal Historical Society, Fifth Series, Vol. 11 (1961), pp. 1–21.

5. Kuehn, p. 8.

6. Ibid., p. 1.

7. John Hayes and John Hattendorf, eds. *The Writings of Stephen B. Luce*. Newport, RI: Naval War College Press, 1975, pp. 39–40.

8. Kuehn, p. 162.

9. Peter Perla, *The Art of Wargaming*. Annapolis, MD: Naval Institute Press, 1990, pp. 72–74.

10. Hattendorf, p. 161.

11. Henry Beers, "The Development of the Office of the Chief of Naval Operations, Part II." *Military Affairs*. American Military Institute. Washington, DC. Fall 1946. pp. 17–18.

12. Albert A. Nofi, *To Train the Fleet for War: The U.S. Navy Fleet Problems, 1923–1940*. Newport, RI: Naval War College Press, 2010, p. xxvi.

13. Ibid., p. 1.

14. Ibid., p. 4.

15. Ibid., pp. 20–21.

16. Ibid., p. 21.

17. Ibid., p. 275.

18. Ibid., p. 294.

19. Ibid., p. 286.

20. Ibid., p. 288.

21. Ibid., p. 292.

22. Ibid., p. 311.

23. Ibid., pp. 159–60.

24. Ibid., p. 303.

25. Ibid., p. 310.

26. Ibid., p. 40.

27. Ibid., p. 41.

28. Ibid., p. 2.

29. Ibid., p. 313.

30. Ibid., p. 314.

31. Ibid., p. 314.

32. Ibid., p. 282.

33. Ibid., p. 319.

34. Ibid., p. 321.
35. Hayes, et al., p. 41.
36. Edward S.Miller, *War Plan Orange: The U.S. Strategy to Defeat Japan, 1897–1945*. Annapolis, MD: Naval Institute Press, 1991, pp. 32–33.
37. Miller, pp. 351–352.
38. John A. Butler, *Sailing on Friday: The Perilous Voyage of America's Merchant Marine*. Washington, DC: Brassey's, 1997, pp. 164–165.
39. Andrew Gibson and Arthur Donovan, *The Abandoned Ocean: A History of United States Maritime Policy*. Columbia, SC: University of South Carolina Press, 2000, pp. 144–145.
40. Alex Roland, et al., *The Way of the Ship: America's Maritime History Reenvisioned, 1600–2000*. Hoboken, NJ: John Wiley & Sons, 2008, p. 299.
41. Gibson and Donovan, pp. 166–167.
42. Roland, p. 372.
43. Ibid., pp. 302–303.
44. Ibid., pp. 319–320.
45. Ibid., pp. 308–309.
46. Miller, p. 145.
47. Ibid., p. 25.
48. Ibid.
49. Ibid., p. 187, emphasis added.
50. Ibid., p. 29.
51. Bradley A. Fiske, *The Navy as a Fighting Machine*. Annapolis, MD: U.S. Naval Institute Press, 1916, pp. 326.
52. Miller, p. 32.
53. John A. Adams, *If Mahan Ran the Great Pacific War*. Bloomington: University of Indiana Press, 2008, p. 28.
54. Ibid., p. 26.
55. Ibid., p. 32.
56. Ibid.
57. Ibid., p. 38, emphasis original.
58. Ibid., p. 130.
59. Ibid.
60. Ibid., p. 142.
61. Ibid., p. 65.
62. Ibid., pp. 178–179.
63. *United Nations Treaties and Principles on Outer Space*. New York: United Nations, 2002, p. 4.
64. Ibid., p. 4.
65. Ibid., p. 31.
66. Ibid.
67. Kuehn, p. 176.
68. Grossnick, pp. 489–506.
69. Dudley W. Knox, *The Eclipse of American Sea Power*. New York: American Army and Navy Journal, 1922, pp. 35–36.
70. Nofi, p. 320.

Chapter 5

1. Peter Schwartz, *The Art of the Long View*. New York: Currency Doubleday, 1991, pp. 3–4.
2. Kees Van Der Heijden, *Scenarios: The Art of Strategic Conversation*. 2nd ed. Hoboken, NJ: John Wiley & Sons, 2005, p. 4.
3. Van Der Heijden, pp. 26–27.
4. Schwartz, p. 19.
5. Ibid., p. 247–8.
6. *Report of the Commission to Assess United States National Security Space Management and Organization*. Washington, DC: Government Printing Office, 11 January 2001, p. 22.
7. Everett C. Dolman, *Astropolitik: Classical Geopolitics in the Space Age*. London: Frank Cass, 2002, pp. 156–158.
8. Peter Garretson and Douglas Kaupa, "Planetary Defense: Potentail Mitigation Roles of the Department of Defense." *Air & Space Power Journal*. Maxwell AFB, AL: Air University, Fall 2008. http://www.airpower.au.af.mil/airchronicles/apj/apj08/fal08/garretson.html (accessed 20 February 2014).
9. Duncan Steel, *Target Earth: The Search for Rogue Asteroids and Doomsday Comets That Threaten Our Planet*. Pleasantville, NY: Reader's Digest, 2000, pp. 110–111.
10. "LINEAR." http://www.ll.mit.edu/mission/space/linear (accessed 20 February 2014).
11. Garretson and Kaupa.
12. Steel, p. 131.
13. Garretson and Kaupa.
14. Travis S. Taylor, *A New American Space Plan*. Riverdale, NY: Baen, 2012, p. 177.
15. William E. Burrows, *The Survival Imperative*. New York: Forge, 2006, pp. 247–248.
16. Colin Gray and John Shelton, "Space Power and the Revolution in Military Affairs: A Glass Half Full?" *Airpower Journal*. Fall 1999. http://www.airpower.maxwell.af.mil/airchronicles/apj/apj99/fal99/gray.html (accessed 20 February 2014).
17. Michael A.G. Michaud, *Contact with Alien Civilizations*. New York: Copernicus Books, 2007, pp. 312–313.
18. Michaud, pp. 375–376.
19. Michael D. Papagiannis, "The Importance of Exploring the Asteroid Belt." *Acta Astronautica*. Vol. 10, No. 10 (1983), p. 711.
20. Travis Taylor and Bob Boan, *Alien Invasion*. Riverdale, NY: Baen, 2011, pp. 180–181.
21. Taylor and Boan, p. 181.
22. Schwartz, p. 9.
23. Van Der Heijden, pp. 36–37.
24. James Dewar, *The Nuclear Rocket: Mak-*

ing the Planet Green, Peaceful, and Prosperous. Burlington, ON: Apogee Books, 2009, p. 37.

25. Alfred T. Mahan, *The Influence of Sea Power upon History, 1600–1783.* Cambridge, MA: John Wilson and Sons, 1890. p. 23.

26. James Holmes, and Toshi Yoshihara, "Mahan's Lingering Ghost." *U.S. Naval Institute Proceedings*, December 2009. Annapolis, MD: Naval Institute Press, 2009, p. 43.

27. Holmes and Yoshihara, p. 43.

28. Toshi Yoshihara and James Holmes, *Red Star over the Pacific: China's Rise and the Challenges to U.S. Maritime Strategy.* Annapolis, MD: Naval Institute Press, 2010, p. 9.

29. Stefan Possony, Jerry Pournelle, and Francis Kane, *The Strategy of Technology.* 1997 Revision (electronic). Chapter 1. http://www.jerrypournelle.com/slowchange/Strat.html (accessed 22 September 2013), p. 13.

30. Ibid., p. 10.

31. Ibid., pp. 12–13.

32. Ibid., p. 12.

33. Ibid., pp. 17–18.

34. Ibid., pp. 79–82.

35. John J. Klein, *Space Warfare: Strategy, Principles, and Policy.* New York: Routledge, 2006, pp. 162–163.

36. Ibid., p. 52.

37. Albert A. Nofi, *To Train the Fleet for War: The U.S. Navy Fleet Problems, 1923–1940.* Newport, RI: Naval War College Press, 2010, p. 319.

38. Taylor, pp. 188–189.

39. Daniel O. Graham, *Confessions of a Cold Warrior.* Fairfax, VA: Preview Press, 1995, pp. 226–227.

40. Graham, p. 216.

41. Lee Valentine, "A Space Roadmap." http://ssi.org/reading/papers/space-studies-institute-roadmap/ (accessed 20 February 2014).

Bibliography

Adams, John A. *If Mahan Ran the Great Pacific War*. Bloomington: University of Indiana Press, 2008.

Adkisson, Richard V. "The Original Institutionalist Perspective on Economy and Its Place in a Pluralistic Paradigm." *International Journal of Pluralism and Economics Education*. Vol. 1, No. 4 (2010).

Baier, Frederick. *50 Questions Every Airman Can Answer*. Maxwell AFB, AL: Air University Press, 1999.

Beers, Henry. "The Development of the Office of the Chief of Naval Operations, Part II." *Military Affairs*. American Military Institute. Washington, DC. Fall 1946.

Bigelow, Robert. Speech given in Las Cruces, NM, 19 October 2011. http://bigelowaerospace.com/The_New_China_Syndrome.pdf (accessed 20 February 2014).

Braynard, Frank O. *S.S. Savannah: The Elegant Steamship*. Atlanta: University of Georgia Press, 1963. Reprint, Garden City, NY: Dover, 1988.

Burrows, William E. *The Survival Imperative*. New York: Forge, 2006.

Butler, John A. *Sailing on Friday: The Perilous Voyage of America's Merchant Marine*. Washington, DC: Brassey's, 1997.

Clausewitz, Carl von. *On War*. Howards and Paret, eds. New York: Knopf, 1999.

Collins, John M. *Military Strategy*. Washington, DC: Potomac Books, 2008.

Dewar, James. *The Nuclear Rocket: Making the Planet Green, Peaceful, and Prosperous*. Ontario: Apogee Books, 2009.

Dolman, Everett C. *Astropolitik: Classical Geopolitics in the Space Age*. London: Frank Cass, 2002.
_____. *Pure Strategy*. London: Frank Cass, 2005.

Dyson, George. "Project Orion: Deep Space Force." 4 March 2008. http://makezine.com/magazine/make-13/project-orion-deep-space-force/ (accessed 20 February 2014).
_____. *Project Orion: The True Story of the Atomic Spaceship*. New York: Owl Books, 2002.

Erickson, Mark. *Into the Unknown Together*. Maxwell AFB, AL: Air University Press, 2005.

Fiske, Bradley A. *The Navy as a Fighting Machine*. Annapolis, MD: U.S. Naval Institute Press, 1916.

Fletcher, Ian. *Free Trade Doesn't Work*. Sheffield, MA: Coalition for a Prosperous America, 2011.

Foster, Richard N. *Innovation: The Attacker's Advantage*. New York: Summit Books, 1986.

Futron Corporation. *Futron's 2012 Space Competitiveness Index*. Bethesda, MD: Futron, 2012.

Garretson, Peter, and Douglas Kaupa. "Planetary Defense: Potential Mitigation Roles of the Department of Defense." *Air & Space Power Journal*. Maxwell AFB, AL: Air University, Fall 2008. http://www.airpower.au.af.mil/airchronicles/apj/apj08/fal08/garretson.html (accessed 20 February 2014).

Gibson, Andrew, and Arthur Donovan. *The Abandoned Ocean: A History of United States Maritime Policy*. Columbia: University of South Carolina Press, 2000.

Godwin, Robert, ed. *Dyna-Soar: Hypersonic Strategic Weapons System.* Burlington, ON: Apogee Press, 2003.

Goetzman, William H. *Exploration and Empire.* New York: History Book Club, 1966.

Gomory, Ralph, and William Baumol. *Global Trade and Conflicting National Interests.* Cambridge, MA: MIT Press, 2000.

Graham, Daniel O. *Confessions of a Cold Warrior.* Fairfax, VA: Preview Press, 1995.

Gray, Colin, and John Shelton. "Space Power and the Revolution in Military Affairs: A Glass Half Full?" *Airpower Journal.* Fall 1999. http://www.airpower.maxwell.af.mil/airchronicles/apj/apj99/fal99/gray.html (accessed 20 February 2014).

Gray, Colin S. "The American Way of War." In McIvor, ed., *Rethinking the Principles of War.* Annapolis, MD: Naval Institute Press, 2005.

Grenville, J.A.S. *Diplomacy and War Plans in the United States, 1890–1917.* Transactions of the Royal Historical Society, Fifth Series, Vol. 11. 1961.

Grossnick, Roy A. *United States Naval Aviation, 1910–1995.* 4th ed. Washington, DC: Naval Historical Center, 1997.

Handel, Michael I. *Masters of War.* Third Revised Edition. New York: Routledge, 2001.

Hattendorf, John B., et al. *Sailors and Scholars: A Centennial History of the United States Naval War College.* Newport, RI: Naval War College Press, 1984.

Hayes, John, and John Hattendorf, eds. *The Writings of Stephen B. Luce.* Newport, RI: Naval War College Press, 1975.

Holland, William J., Jr. "Strategy and Submarine." *United States Naval Institute Proceedings.* December 2013.

Holley, I.B. *Ideas and Weapons.* New Haven, CT: Yale University Press, 1953. Reprint, Washington, DC: Air Force History and Museums Program, 1997.

_____. *Technology and Military Doctrine.* Maxwell AFB, AL: Air University Press, 2004.

Holmes, James, and Yoshihara, Toshi. "Mahan's Lingering Ghost." *U.S. Naval Institute Proceedings,* December 2009. Annapolis, MD: Naval Institute Press, 2009.

Johnson-Freese, Joan. *Space as a Strategic Asset.* New York: Columbia University Press, 2007.

_____. "Will China Overtake America in Space?" 20 June 2012. http://www.cnn.com/2012/06/20/opinion/freese-china-space/ (accessed 20 February 2014).

Knox, Dudley W. *The Eclipse of American Sea Power.* New York: American Army and Navy Journal, 1922.

Kuehn, John T. *Agents of Innovation: The General Board and the Design of the Fleet That Defeated the Japanese Navy.* Annapolis, MD: Naval Institute Press, 2008.

Lewis, John S. *Mining the Sky.* Reading, MA: Addison-Wesley, 1996.

"LINEAR." http://www.ll.mit.edu/mission/space/linear (accessed 20 February 2014).

Long, K.F. *Deep Space Propulsion.* New York: Springer, 2012.

Lupton, David. *On Space Warfare: A Space Power Doctrine.* Maxwell AFB, AL: Air University Press, 1988.

Mahan, Alfred T. *The Influence of Sea Power Upon History, 1600–1783.* Cambridge, MA: John Wilson and Sons, 1890.

_____. *Retrospect and Prospect.* Boston, MA: Little, Brown, 1902.

McCray, W. Patrick. *The Visioneers.* Princeton: Princeton University Press. 2013.

Meilinger, Philip S. *10 Propositions Regarding Air Power.* Washington, DC: Air Force History and Museums Program, 1995.

Michaud, Michael A.G. *Contact with Alien Civilizations.* New York: Copernicus Books, 2007.

Miller, Edward S. *War Plan Orange: The U.S. Strategy to Defeat Japan, 1897–1945.* Annapolis, MD: Naval Institute Press, 1991.

Mitchell, William. *Winged Defense.* In Jablonsky, ed., *Roots of Strategy, Book Four.* Mechanicsburg, PA: Stackpole Books, 1999.

Nofi, Albert A. *To Train the Fleet for War: The U.S. Navy Fleet Problems, 1923–1940.* Newport, RI: Naval War College Press, 2010.

Nye, Joseph S. *Soft Power.* New York: Public Affairs, 2004.

Oberg, James E. *Space Power Theory.* Colorado Springs, CO: United States Space Command, 1999.

Okuda, Michael, and Denise Okuda. *The Star Trek Encyclopedia.* Revised Edition. New York: Pocket Books, 1999.

Papagiannis, Michael D. "The Importance of Exploring the Asteroid Belt." *Acta Astronautica.* Vol. 10, No. 10, 1983.

Perla, Peter. *The Art of Wargaming.* Annapolis, MD: Naval Institute Press, 1990.

Philbrick, Nathaniel. *Sea of Glory: America's Voyage of Discovery, The U.S. Exploring Expedition, 1838–1842.* New York: Viking Press, 2003.

Porter, Michael. *The Competitive Advantage of Nations.* New York: Free Press, 1990.

Possony, Stefan T., and Jerry Pournelle. *The Strategy of Technology.* Cambridge, MA: Dunellen, 1970.

_____, _____, and Francis Kane. *The Strategy of Technology.* 1997 Revision (electronic). Chapter 1. http://www.jerrypournelle.com/slowchange/Strat.html (accessed 22 September 2013).

Project Horizon: A U.S. Army Study for the Establishment of a Lunar Military Outpost. Volume I. U.S. Army Ballistics Missile Agency. 8 June 1959. http://www.history.army.mil/faq/horizon/Horizon_V1.pdf (accessed 20 February 2014).

Radlauer, Edward, and Ruth Radlauer. *Atoms Afloat! The Nuclear Ship* Savannah. London: Abelard-Schuman, 1963.

Report of the Commission to Assess United States National Security Space Management and Organization. Washington, DC: Government Printing Office, 11 January 2001.

Rickards, James. *Currency Wars: The Making of the Next Global Crisis.* New York: Penguin, 2011.

Roland, Alex, et al. *The Way of the Ship: America's Maritime History Reenvisioned, 1600–2000.* Hoboken, NJ: John Wiley & Sons, 2008.

Ronda, James R. *Beyond Lewis & Clark: The Army Explores the West.* Tacoma: Washington State Historical Society, 2003.

Schumpeter, Joseph. *The Theory of Economic Development.* Cambridge, MA: Harvard University Press, 1934. Reprint, New Brunswick, NJ: Transaction Press, 2008.

Schwartz, Peter. *The Art of the Long View.* New York: Currency Doubleday, 1991.

Seedhouse, Erik. *The New Space Race: China vs. the United States.* Chichester, UK: Praxis, 2010.

Sellers, Jerry. *Understanding Space: An Introduction to Astronautics.* Revised Second Edition. Boston: McGraw-Hill, 2004.

Shufeldt, Robert W. *The Relation of the Navy to the Commerce of the United States.* Washington, DC: John L. Gink, 1878.

Shulman, Mark R. *Navalism and the Emergence of American Sea Power, 1882–1893.* Annapolis, MD: Naval Institute Press, 1995.

Space Foundation. *The Space Report 2013.* Colorado Springs, CO: Space Foundation, 2013.

Steel, Duncan. *Target Earth: The Search for Rogue Asteroids and Doomsday Comets That Threaten Our Planet.* Pleasantville, NY: Reader's Digest, 2000.

Taylor, Travis S. *A New American Space Plan.* Riverdale, NY: Baen, 2012.

_____, and Bob Boan. *Alien Invasion.* Riverdale, NY: Baen, 2011.

Tirpak, John. "The Space Commission Reports." *AIR FORCE Magazine.* March 2001.

Tkacik, John J. "China Space Program Shoots for Moon." *The Washington Times,* 8 January 2010. http://www.washingtontimes.com/news/2010/jan/08/china-eyes-high-ground/?page=2#ixzz2czkK0BhW (accessed 20 February 2014).

United Nations Treaties and Principles on Outer Space. New York: United Nations, 2002.

Valentine, Lee. "A Space Roadmap." http://ssi.org/reading/papers/space-studies-institute-roadmap/ (accessed 20 February 2014).

Van Der Heijden, Kees. *Scenarios: The Art of Strategic Conversation* 2nd ed. Hoboken, NJ: John Wiley & Sons, 2005.

Von Braun, Wernher. *The Mars Project.* Trans. Henry White. Urbana: University of Illinois Press, 1953.

Ward, Susan. "Business Plan." http://sbinfocanada.about.com/cs/startup/g/businessplan.htm. (accessed 20 February 2014).

Wingo, Dennis. "Economic Development of the Solar System: The Heart of a 21st-Century Spacepower Theory." In Lutes, et al., eds. *Toward a Theory of Spacepower: Selected Essays.* http://www.ndu.edu/press/lib/pdf/spacepower/spacepower.pdf (accessed 22 February 2014).

Winton, Harold. "On the Nature of Military Theory." In Lutes, et al., eds. *Toward a Theory of Spacepower 2*. Washington, DC: National Defense University Press, 2011.

Worden, Simon P., and John E. Shaw. *Whither Space Power? Forging a Strategy for the New Century*. Maxwell AFB, AL: Air University Press, 2002

Wylie, J.C. *Military Strategy: A General Theory of Power Control*. Annapolis, MD: Naval Institute Press, 1967.

Yoshihara, Toshi, and James Holmes. *Red Star Over the Pacific: China's Rise and the Challenges to U.S. Maritime Strategy*. Annapolis, MD: Naval Institute Press, 2010.

Index

Page numbers in **bold italics** indicate pages with illustrations.

ability (space power) 12–15, 28–40, 45–48, 92, 102, 104, 109–115, 149, 158, 214, 218–219, 227, 230, 232; definition of 32; in Space Power Logic Delta 23, *25*, 26, *27*
absolute value 4
access (space power) 22–27, 30–34, *35*, 48, 55–60, 71, 76, 92, 104, 106, 109–115, 137–138, 149, 158, 194, 196, 219, 229–232; definition of 23; in Space Power Grammar Delta 15, *16*, 17, *18*
Access Technology 114–115, 137
Advanced Space Operations School (ASOpS) 236–237
Agents of Innovation 188
aggressor 121–128
Air Corps Tactical School 64–65, 104, 235
Air Force Astronautics Division 80
Air Force Space Command 11, 14, 25, 74, 76–78, 156, 205, 237–238
Air Force Special Weapons Center 94
Air Force Systems Command 101
Air Force, U.S. 11–12, 37, 47, 76–81, 84–96, 100–102, 145, 235, 237
Air Force Weapons School Space Division 178
air power 11–13, 90–91, 99–103, 235
Air University 74, 207, 235–236
aircraft carrier 111, 154, 204, 215; and Fleet Problems 170–175, 179, 187; interwar naval revolution 160–161, 195–198; Washington Naval Treaty and 191
airships 169, 174–175
Akron (ZRS 4) 175
alien invasion 216–217, 222
Alpha Centauri 147
altitude 20, *35*, 36–37, 75, 92
American Astronautical Society 241
American Expeditionary Force 99
Ansari X-PRIZE 137

antimatter 45, 139, 141–142
antimatter rocket 142
antisatellite weapon (ASAT) 92, 206, 208
Apollo Program 29, 49, 59, 67–69, 71–72, 80, 93, 96, 230, 239
Aristotelian view of planetary defense 218–220; *see also* planetary defense
Army Ballistic Missile Agency 80
Army, U.S. 47, 67, 73, 76–83, 90, 99–101, 171, 187, 208, 239
Arnold, General Henry "Hap," USA 99
asteroid 20, 29, 37, 48, 70–72, 84, 95, 143, 212–217, 220–222
Astropolitik 74, 208–209
Atlas rocket 225
Atomic Energy Commission (AEC) 137, 148
Atoms for Peace 151
auxiliary ships 161, 168, 173, 190
aviation 99, 238; *see also* naval aviation
Aviation Week and Space Technology 234

B-1 bomber 145
B612 Foundation 213
Baier, Captain Fritz, USAF 243
bases 15, 22, 55–58, 92, 222, 228–229
Battlestar Galactica 226
Baumol, William 134–135
Beers, Henry 167
Beidou (Chinese satellite navigation system) 2
Bigelow, Robert 3
Bigelow Aerospace 46
Boeing 46, 75, 130
Boushey, Brigadier General Homer, USAF 79–80, 92
British Empire 188, 228
British Interplanetary Society 142
British Royal Customs Service 153

Buck Rogers 79
Bureau of Aeronatics 161, 164, 197
Bureau of Navigation 161, 164
bureau system, U.S. Navy 161–162, *163*, 164, 168
Burnett-Stuart, Sir John 65
Burrows, William 215
business plan 28–30, 55, 195, 198

Cadet Corps, Merchant Marine 182–183, 187
California Institute of Technology 79
Cameron, James 29
capabilities based development 147–149
carrier *see* aircraft carrier
Cavalry, U.S. Army 100–101
Central Intelligence Agency (CIA) 12
Ceremonial values *see* values
chemical (deposits) 37, 45
chemical rocket 49, 93, 111, 114, 137–141, 152, 154, 206, 224–227
Chief of Naval Operations (OpNav) 162, *163*, 166–169, 171, 191, 194–195
China 2–5, 7, 83, 107, 136, 177, 188, 190
Civil War, U.S. 130, 193
Clarke, Arthur C. 73, 92
Clausewitz, Carl von 14, 24, 31, 116, 119–121, 124–125
Clementine I mission 84
Coast Guard, U.S. 102, 143, 153
Cochrane, Zephram 147
Cold War 10, 29, 66, 73–74, 76, 94, 107, 145, 172, 179, 234
Collins, Colonel John M., USA 115
Colonies (space power) 19–22, 49, 55–58, 71, 92, 106, 221, 228–229; definition of 21; in the Grammar of Space Power 15, *16*, 17, *18*, 31, 34, *35*, in space power development 39, *41*, 45–46
Colorado Springs, Colorado 82, 236
Combinations (space power) *18*, 22–23, 26, 36, 110, 113, 121, 164, 195–196, 232; definition of 22; in organizational development 62–63; role in space development 39, *41*, 42–43, 46
comet 95, 210, 212, 215, 220
Commander in Chief United States Fleet (CINCUS.) 168, 170–171, 175–176
commerce 15, 17, 21–26, 34, 85, 150, 190–193, 208, 229–231, 236; heresy of commerce 58–60; role in merchant marine 180, 182–185; role of navy in commerce 131–133; sub-grammar of commerce *55*, 56–58
commerce heresy 58–60
Congress, U.S. 77, 79, 163, 174, 180, 182
conservator 121–128
Constellation Program 77
Contact 69
Coontz, Admiral Robert, USN 170
Copernican view of planetary defense 218–220; *see also* planetary defense
Corps of Topographical Engineers 82

culminating point 119, *120*, 121, *122*, 124–125, 127–128
Cunningham, A.C. 191

Deep Space Force 92–96
deflection dilemma 214
Delta Clipper (DC-X) 73, 239
Department of Defense (DOD) 76, 84, 211, 214
development paths 58, 62–63, 97–98, 102–103, 106–107, 110, 112, 194–200, 234, 238, 240; space power paths 39–40, *41*, *42*, 43–47
Diplomatic, Informational, Military, Economic (DIME) power 31, 81
Director of Naval Intelligence 164; *see also* Office of Naval Intelligence
Dr. Strangelove 95
doctrine 11, 28, 195, 198, 209, 235; definition of 30; development of in Navy 169–172, 177–178; Holley on space doctrine 98–104, 121, 129; importance of for military development 63–65, 89; as a space power transformer 30, 55, 58
Dolman, Everett 47–48, 74, 207, 209–210
Douhet, Giulio 90–91
HMS *Dreadnought* 153, 169, 231
Dresden 91
DuBridge, Lee 79–80
Dunnigan, James F. 200
Dyna-Soar 88
Dyson, Freeman 93
Dyson, George 93, 95–96

Earth 1, 21, 23, 36–38, 45–49, 67–71, 75, 79–80, 84, 88, 94–95, 111, 138–139, 143–144, 147, 152, 180, 201, 204, 207–223, 228–229, 241
Eat at Joe's (scenario) 216
Eclipse of American Sea Power 190, 199
economic development 39–47, 106, 110
economic logic 112–114
economic power 17, 47, 50, 55, 60, 76, 81–82, 90, 114, 130–132, 205; definition of 28–29; in Logic of Space Power 23, *25*, 26, *27*, 31–32, *42*, 43–44
An Economic Survey of the American Merchant Marine 183
economic warfare 232, 240–241
economics 7, 10, 17–20, 23–26, 28–29, 34, 39, 41–42, 183–184, 208–209, 231–232; free market vs. mercantilism 129–137; Original Institutionalist school 64–65, 84, 97, 100, 102–103
education 7, 16, 18–19, 135, 235, 238
Ehricke, Krafft A. 214
Eisenhower, Dwight 79–81, 84–90, 96–97, 104, 146, 151
electromagnetic (EM) radiation 17, 21–22, 207, 222
elements (space power) 7, 9–10, 31–34, 38–40, *41*, 42–56, 49–50, 53–58, 62, 98, 106, 108–112, 114, 188–189, 194–196, 205, 232; definition of *16*, 17, *18*; description 19–24

elements of sea power 15, 19–20, 24
enablers (space power) 16–17, *18*, 44–45, 49; definition of 19
Engineering Breakthrough phase 107, 109–110
Enterprise (Star Trek) 149, 153–154, 226
entrepreneur 44, 62–64
Essex class aircraft carrier 198
Europa 219
Europe 5, 21, 83, 151, 170
exclusivity of capabilities/knowledge 18–19
Exploration and Empire 50–51
extraterrestrials 218–219, 221–222

F-15 Eagle 145
F-16 Fighting Falcon 145
F-22 Raptor 214
facilities 18–19, 69, 208
Fail-Safe 95
Falcon I rocket 157
Falcon 9 rocket 157
Fast Carrier Task Force 170, 172–173, 179, 196; *see also* aircraft carrier
Federal Emergency Management Agency (FEMA) 212
Fisk, Admiral Bradley, USN 188
Fleet Base Force 181, 191–2, 198
Fleet Problems, Navy 168–280, 195–298, 200, 238
Fletcher, Ian 130, 136
floating dry dock 160–161, 181, 191, 195, 197–199
Fortification Clause (Article XIX, Washington Naval Treaty) 188–192, 199
Foster, Richard N. 117, 125, 127–128
free market 129, 132, 208–209
free trade 130, 134, 136
Freedom Space Station 29
Fremont, John 82
Fuller, J.F.C 65
fusion 37, 45, 139, 141–142, 154, 221
fusion rocket 142
Futron Corporation 4–5
futurism 6, 11, 38, 142

Gagarin, Yuri 75
Galloway, Eileen 96
Garretson, Lieutenant Colonel Peter, USAF 214
Gemini (U.S. manned spacecraft) 3, 96
General Board, Air Service 237
General Board, Navy 162, *163*, 164, 167–169, 171, 179, role interpreted through General Theory 194–196
General Board, Space 237–238, 240
general staff 162–163, 235–236
General Theory of Space Power 53–55, 58, 62–63, 74, 97, 108–109, 114, 159, 164, 179, 193–195, 200, 232; description 13–16, 19, 28–31; development theory 39–40, 47–51; intent of 9–11; prescriptions of 102–105
Geneva Convention 216

geography 18–19, 166, 170
giggle factor 211–212
Global Positioning System (GPS) 1, 22, 37, 77, 205
Global Trade and Conflicting National Interests 134
global utilities 10, 84
Goetzmann, William 50–51
Goldwater Nichols Defense Reorganization Act 155–156
Gomory, Ralph 134
Gomory/Baumol trade theory 134–135
Graham, Lieutenant General Daniel, USA 66, 72–74, 81–82, 97, 239–241
Grammar Delta 11, 22, 26, 31–33, 61, 106, 108–111, 114, 137, 155, 196; definition of 14–18; and space power development 39–43, 45–46, 49–53
grammar of commerce *55*, 57, 60, 193
grammar of politics *55*, 56, 58, 190, 192–193
Grammar of Space Power 7, 11, 31–33, 60–61, 99–100, 102, 104–105, 106–115, 158, 193, 240–241; definition of 16; description of 14–15, *16*, 17, *18*, 22–23, 26; development through 39–51, sub-grammars 53–58
grammar of war 23, *55*, 56–59, 188–193
Gray, Colin (strategist) 11, 216–217
Great Depression 178
Grenville, J.A.S. 163
Grimes, James M. 173
Ground-based Electro-Optical Deep Space Surveillance (GEODSS) 211
growth 4–6, 10, 28, 59, 99, 106, 132

Hamburg 91
Hammer of God (scenario) 210
Handel, Michael E. 119–125
Handel's Model of Conflict *120*, 121–125
hardware 16–19, 40, 63, 66, 101, 194
Hawaii 180, 189–190, 199, 231
Helium-3 37, 142
Heppenheimer, T.A. 75
Heresies of space power 57–60, 193
Hierarchy of space power *see* space exploitation
High Frontier: A New National Security Strategy 73
High Frontier: Human Colonies in Space 71
Hiroshima 91
Holley, Major General I.B., USAFR 30, 63–65, 99–104, 121, 129
Holmes, James 14–15, 23, 180, 228
Homer, Richard 93
Hubble Space Telescope 49
Hunley, H.L. 87

Imperial Japanese Navy (IJN) 177, 189–191
imperialism 210; interstellar 218
impulse drive 75, 137, 141–142, 154–155
industry 5, 52, 59, 71, 78, 108, 111, 133–137, 148,

157, 194; in development of space power 39–
43, 46; as enabler of space power 18–19
Influence of Sea Power Upon History 87, 96
Influence Technology 114–115
innovation 6, 28, 188, 191–200, 211, 226–227,
238, 240; consumer versus producer 155–157;
entrepreneurs and 60, 63–65; navalist innovation 161–166, 169–178; spillovers in development 41–43, 46, 103, 108, 110, 116–117, 121,
126–128, subsidizing 137–138
Innovation: The Attacker's Advantage 117
instrumental values *see* values
Intellectual Breakthrough phase 50, 107–108
International Space Station (ISS) 22, 29, 45, 58–
59, 68, 76
interstellar travel 6, 10, 53, 70, 111, 147, 217–218,
225
Invention Breakthrough phase 107–108

Janus 78–79
Japan 5, 136; in World War II 160–161, 165–170,
172, 174, 177, 180–181, 184, 187–191, 199, 201
Jefferson, Thomas 82
Jeremiah, Admiral David, USN 98
Jet Propulsion Laboratory 69
Johnson, Lyndon B. 97
Johnson-Freese, Joan 29
Jupiter 31–32, 75, 114, 218
USS *Jupiter* 198
Jutland, Battle of 169

Kane, Colonel Frank X., USAF 232
Kennedy, John F. 97, 146
Kennedy, Joseph P. 182
Killian, James 79–80, 85, 96
Kings Point, NY *see* United States Merchant
Marine Academy
Kirtland Air Force Base, NM 94
Kistiakowsky, George 87, 96
Kitty Hawk 87, 235
Klein, Commander John, USN 236
Klingon 143
knowledge spillovers 41–43, 58
Knox, Captain Dudley W., USN 189–190, 199
Korea 130
Kuehn, Commander John T., USN 162–164,
188, 191
Kuiper Belt 217

Land, Rear Admiral Emory S., USNR 182–183
USS *Langley* 198
launch 71, 73, 77, 94, 109, 136–137, 156–157,
206–207, 209–210, 213–216, 219–220, 222,
224–226, 229, 239–240; and space power ranking 1–2, 5–6
levels of space exploitation *see* space exploitation
Lewis, John S. 49–50, 72
Lewis and Clark expedition 82
Liberty engine 145

Liberty ship 181, 185
Liddell Hart, Captain B.H., Royal Army 65
Lincoln Laboratory 211
Lincoln Near-Earth Asteroid Research project
(LINEAR) 212–213
lines of communication 15, 17, 20, 24, 43, 56,
120–121, 213, 217
Lockheed Martin 46, 130
Logic Delta 11, 14–15, 17, *25*, *27*, 61–62, 106,
108, 110, 114, 179, 195, 197; definition of 25–
28, 32–33, 44, and space power development
46–51, 55
Logic of Space Power 7, 14, 16–17, 62–63, 97–98,
102–105, 110–111, 114–115, 158, 240; definition
of 23–24, *25*; description of 26, *27*, 31–34;
development through 39, *42*, 43–44, 46–51,
53–60
Long, John D. 163
Long March (Chinese launch vehicle) 2
Los Angeles (ZR 3 zeppelin) 175
Los Angeles class submarine 148
Luce, Rear Admiral Stephen B., USN 159, 161,
164, 179–181
lunar base *see* moon base
Lupton, David Lieutenant Colonel, USAF 12–13

Macon (ZRS 5) 175
Mahan, Rear Admiral Alfred Thayer, USN 9–10,
importance in naval development 87, influence
on Deep Space Force 96, 130, 159, 165, 188,
191–193, peaceful strategic offensive 228–229,
231; sea power theory of 14–15, 18–19, 23–24,
31, 34
Mahan's Lingering Ghost 24
Management Breakthrough phase 107–109
Marine Corps, U.S. 23, 47, 164
Maritime Commission, U.S. (MarComm) 182–
185
maritime operational access 23–24
markets 15, and colonies 17, 56–57; and elements
of space power 20–21, 24, 26, as path development 39, 43–45, 194, 109, 136, 148, and peaceful strategic offensive 228–230
Mars 3, 21, 33, 38, 49, 52, 59, 70, 92–93, 95, 111,
141, 143, 149, 157, 197, 218–219, 222, 229–230;
and Von Braun vision 67–68
Mars Project 67
Mars Society 54
Marshall Space Flight Center 80
Matloff, Gregory 221
Maury, Matthew F. 131
McNulty, Rear Admiral Richard M., USN 182–
183
Meilinger, Colonel Philip, USAF 12
mercantilism 129–132, 134
merchant marine 15, 21, 57, 154, 198–200; development of 130–134; development interpreted
through General Theory 179–190
Merchant Marine Act of 1920 184

Merchant Marine Act of 1936 174, 187, 198
Mercury 88, 96
Michaud, Michael 218, 220
microsatellite 6, 156, 222, 228
militarization of space 60
military power 10, 12, 14, 50, 54–55, 60, 76–77, 81–82, 88, 98, 116, 195, 198, 229, 232; development of *42*, 43–44, 46–47; in Logic of Space Power 23–24, *25*, *27*, 28–35; *see also* space power
military space commands 76–78
Mining the Sky 49, 72
mission-based development 146, 148
mission pull 148
missionitis 144, 146–149, 152
Mitchell, Brigadier General Billy, USA 11–13, 90–91, 100, 102
Mixson, Captain Donald, USAF 94–96
Moon 3, 10, 21, 37–38, 49, 51, 68–69, 72, 75, 79–81, 84, 92–93, 95, 111, 132, 143–144, 146, 152, 192–193, 212, 215–216, 218, 223, 227, 230
moon base 11, 38, 54, 80–81, 92
Moon Treaty of 1979 193
Musk, Elon 157

N-Squared rule 180
Nagasaki 91
National Advisory Council on Aeronautics (NACA) 85
National Aeronautics and Space Administration (NASA) 11, 45, 49, 51–53, 93, 96–97, 104, 113, 137, 140, 144, 149, 211–214, 230, 239; and Von Braun vision 67–69, 71; schism with military space effort 76–80, 84–89
National Science Foundation 87
National Security Space Institute (NSSI) 236–237
National Space Society 54
natural resources *see* resources
Nautilus submarine 148, 150
naval aviation 111, 164, 169, 172–173, 196–297; *see also* aviation
naval bases 56, 120–121, 161, 165, 168, 180–181, 188–192, 198
Naval War College, United States 3, 104; role in sea power development 162, *163*, 164–169, 171–172, 176, 179, 194–196, 235
navalists 159–169, 179, 183–185, 193–195, 197–201, 231, 234–235, 238
navies 15, 130, 135, 153, 179, 208, 215
Navy, U.S. 7, 15, 24, 47, 68, 76–78, 226, 228, 236–238; development between World Wars 159–183; role in commerce development 130–132, 148, 153; role in developing merchant marine 185–198; role in exploration 82–83, 86, 94
Near Earth Asteroid Tracking team (NEAT) 212
Near Earth Object (NEO) 212, 214–215
Nemesis (hypothetical) 217–219

A New American Space Program 238
New Worlds *35*, 38, 106, 138
New York 41, 150, 187, 239
Nimitz, Fleet Admiral Chester, USN 165–166
Nofi, Albert 168–170, 172, 174–175, 177, 179, 200, 238
North Atlantic Treaty Organization (NATO) 95, 213
nuclear continuum 139, *140*, 142, 149, 152, 154
Nuclear Engine for Rocket Vehicle Application (NERVA) 141, 225–226
Nuclear Hammer 137–138, 224, 228
Nuclear Pulse Propulsion (NPP) 225–227; *see also* Project Orion
nuclear rocket 45, 49, 111, 137–139, 142, 149, 152, 154–155, 225
The Nuclear Rocket 137
Nuclear Thermal Rocket (NTR) *see* nuclear rocket
Nye, Joseph 29

Oberg, Jim 18–19, 208
Office of Naval Intelligence 167
officer 11–12, 58, 63–65, 73–74, 79, 81–83, 86, 88, 93, 95, 100, 103–104, 130, 132, 152, 182, 186–189, 194, 198, 200; naval officers 158–169, 171, 174–177; space officers 231–240; *see also* staff officer
O'Neill, Gerard 66, 71–2, 74, 81–82, 85, 89–90, 97
Operation Highjump 68
operational level of war 15, 33, 115, 162, 168, 177–178, 192
Operationally Responsive Space (ORS) 156
orbital mechanics 20, *35*, 36–37, 111, 150
Original Institutionalist School 64
Orion (nuclear space ship) 92–96, 139, 226–227
Outer Space Treaty of 1967 59, 192–193, 209

Pacific War (World War II) 7, 10, 160–162, 165–174, 179–181, 185–188, 191, 193, 196–200, 201, 234
Pale Blue Dot 69
Panama Canal 83, 180
Pappagiannis, Michael 221
Paths *see* development paths
Pax Americana 204
peaceful strategic offensive 34, 56, 228–232
Pearl Harbor 166, 172–174, 186, 188
Perla, Peter 165
Peterson Air Force Base, Colorado 237
Philbrick, Nathaniel 83
Philippines 170, 180, 188–189, 201, 231
Pike, Zebulon 82
Pioneer 70
planetary defense 48, 70, 143, 220–223; grammar of 204, 210–216; logic of 204, 216–220
Planetary Resources (company) 29

Planetary Society 69
Pluto 70, 154
Plutoid 219
Polaris Submarine 94
policy 10–11, 15, 50–52, 58–60, 86, 96–97, 104, 133, 137, 158, 160, 162, 164, 179, 183, 188–190, 194, 197, 204, 212, 236
political heresy 58–59, 192–193, 199
political power 9–10, 50, 55–56, 60, 76, 82, 104, 180, 195, 205, 230, 240–241; development of *42*, 43–47; in Logic of Space Power 23–24, *25*, 26, *27*, 28–34
populace 18
Porter, Michael 135
Position, Navigation and Timing (PNT) 156; *see also* Global Positioning System
Possony, Stefan 50, 107, 109, 115–118, 124, 129, 232–237
Pournelle, Jerry 50, 73, 107, 109, 115–118, 124, 129, 232–237
power (general) 47–48
Power, General Thomas, USAF 86, 91
Pratt, Admiral William V., USN 175, 194
President's Science Advisory Committee (PSAC) 79
Princeton Institute for Advanced Study 70
principle conditions (space power) *see* enablers (space power)
probability 110, 203–204
production (general) 15, 57–58, 134, 155, 184, 197
production (space power) 15, *18*, 19–24, 29, 31, 34, *35*, 39–40, *41*, 42, 45, 49, 55, 57, 106, 109–113, 194; definition of *16*, 17
Project Horizon 79–81
Project Mars: A Technical Tale 67
Project Orion *see* Orion
Project Orion: The True Story of the Atomic Spaceship 93
Prometheus (fictional ship) 149
Prometheus project 149
pursuit 116, 119–121, *122*, 124

radar 10, 179, 207
Ramey, James T. 148
rate of change 4–5
Reagan, Ronald 29
realist school 11, 73–74
Relation of the Navy to the Commerce of the United States 130
resources 15, 17, 24, 29, 49, 51, 57, 59, 68, 84–85, 95, 106, 112, 138, 144, 193, 209, 230–233; as an enabler of space power 19; as path development 45–46; as production 20; level of space exploitation *35*, 37–38
Return to the Moon 72
revolution 39, 45, 111–112, 159, 161, 169, 191, 194–195, 224, 227–228
Ricardo, David 134

Richardson, Admiral James O., USN 168, 175, 177
Rickards, James 129
rocket 2, 17, 21, 23, 33, 35–36, 38, 45, 49, 66–68, 73, 75, 77, 93, 109, 111, 113, 114, 131, 137, 138, 141–142, 149, 152, 157, 206, 225–227
Rogers, Moses 150
Ronda, James 82–83
Roosevelt, Franklin 174, 178, 182
Roosevelt, Theodore 163
Royal Navy (United Kingdom) 149
Russia 2, 5, 136, 180

S-curve 117, *118*, 121, *122*, 123, 125–128, 139, 149, 224–228
Sagan, Carl 66, 69–72, 80, 85, 90, 96–97
satellite communications 1–2, 28–29, 35, 77, 137, 144, 205
satellite navigation 1–2, 5, 20, 37, 74, 77, 82, 144, 156, 205
satellites 2, 6, 10, 28, 54, 73, 76–77, 80–81, 84, 92, 109, 111, 136–137, 144–145, 156–157, 201, 205–208, 222; as colonies 20–23; in hierarchy of space access 35–39
Saturn 93, 95
Saturn V 67
NS *Savannah* 150–151
SS *Savannah* 149–151
USCGC *Savannah* 149, 152–155
scenario planning 51, 201–205
Schmitt, Harrison 71
Schofield, Admiral Frank, USN 175, 190, 194
School for Advanced Air and Space Studies 236
Schriever, General Bernard, USAF 85–86, 233
Schwartz, Peter 202–203
science fiction 1, 10–11, 20–21, 29, 38, 45, 73–74, 92, 107, 155, 216; role in space development 51–54
science in space power 48–51
Sea of Glory 83
sea power 23–25, 28, 31, 111, 130, 132, 151, 228–229, 231; development of 158–159, 161, 164, 172, 179–199; similarities to space power 7, 10–11, 14–16, 19–20
Seabees 181
Search for Extraterrestrial Intelligence (SETI) 69, 221
Seawolf submarine 148
Secretary of the Navy 161–164, 167–168, 175
Seedhouse, Erik 3
Shaw, John E. 84
Shell Oil Company 202
Shenzhou (Chinese manned spacecraft) 2–3
shipping (general) 15, 57, 133, 150–151, 179, 182–186, 190, 198
shipping (space power) *16*, *18*, 19–23, 29, 31, 34, *35*, 49, 55, 57–58, 106, 194; definition of 17, 21; development of space power through 39–40, *41*, 106, 194

Shufledt, Rear Admiral Robert W., USN 130–134

Shulman, Mark Russell 160

Shumpeter, Joseph 6, 9, 39, 42, 46, 62–63, 110, 112–114, 155–156, 194

Signal Corps, U.S. Army 99–101

soft power 28–30, 55, 59, 195, 199

solar energy *35*, 36–37, 111, 222

solar power satellites 71, 111

Soviet Union 3, 59, 73, 77, 91, 95, 107

Space Academy 237–240

Space Commission (2001) 98

Space Competitiveness Index 4–5

space debris 95, 143, 206–209, 213, 223

Space Education and Training Center 236

space exploitation 34, *35*, 36–38, 71

Space Exploration Technologies (SpaceX) 46, 130, 157, 230

Space Force (hypothetical) 98, 101–102

Space Foundation 2

Space Frontier Foundation 54, 71–72

Space General Board *see* General Board, Space

Space Guard (hypothetical) 102, 143, 152–153

space navigation 206

Space Operational Command 234, 237–238

Space Pearl Harbor (scenario) 204–207, 210

space power 1, 4–7; absolute value measurement 4; definition of 12–14, 23; development of 6–7, 115, 155; Grammar of 7; Logic of 7; paradox of 2, 4–5; rate of change measurement 4; *see also* General Theory of Space Power

Space Report 2013 2

Space Shuttle (U.S. manned spacecraft) 3, 21–22, 68, 71–72, 75–77, 96, 144, 152, 212

space situational awareness (SSA) 207, 211, 213; deep space (SSA-DS) 213–215, 220–221; near Earth (SSA-NE) 207, 209

Space Studies Institute (SSI) 71, 241

Space War College 234–237

space weapons 207, 209, 214–215, 220, 222

Spacer officer 231, 235, 239

Spacer Triad 237–238, 240

Spacers (concept) 202–224, 232, 234, 238, 240; definition of 201

Spacers' War 200–201, 204–205, 223, 232–234

spaceship 93–94, 139, 141, 143, 149–150, 154, 226

Spaceship Company 130, 137

SpaceShipOne 137

Spanish American War 163, 231

Specific Impulse (Isp) 93, 138–139, *140*, 141–142, 226

speed of light 142, 147, 217, 227

Sputnik 1, 36

staff officer 103–104, 235–236; *see also* officer

Star Trek 52–53, 73, 76, 95, 97, 138–142, 147, 153–154

Star Trek Encyclopedia 142

Star Trek Starfleet Technical Manual 52

Star Wars *see* Strategic Defense Initiative (SDI)

Starfleet 72, 89, 152–154

starship 6, 21, 69, 95, 106, 149, 153–154, 227

state capitalism 129

State Department, U.S. 3, 59

steam engine 110–111, 113, 139, 150–151, 159

Steel, Duncan 213–214

steel industry, U.S. 135

Stine, G.H. 75

Stokes, Grant 211–212

strategic access 189–192

Strategic Air Command (SAC) 86, 90–92, 94, 101

strategic culture 89–96, 98–99

Strategic Defense Initiative (SDI) 73, 84, 211

strategic dominance 234

strategic tariff 136

strategy 4, 23–24, 44, 47, 51, 58, 98, 109, 129, 175, 201–203, 209, 211, 213; acquisition 156–158, 160; economic 134–136; ends/ways/means 14–15; entrepreneur's role 63, 65; Fleet Problems' role 196–199; navy development 162–166, 168, 178–193; and planetary defense 216–223; space lines of effort 227, 229, 234–236, 239–240; in space visions 73–74, 78–79, 81, 88–89; of technology 115–124; and theory 6–7, 10

The Strategy of Technology 50, 107, 116, 232

subgrammars of space power 54, *55*, 56–58

Swanson, Claude 168, 175

Swartz, Peter 176

systems dynamics 11

systems engineering 144–146

SzentGyorgi, Albert 224

Taking the High Ground (scenario) 207–210

tariff 136

Taylor, Captain H.C., USN 163

Taylor, Travis 215, 222, 238–239

technological logic 112–114

Technological Process Model 50, 106–110

technological pursuit *see* pursuit

technological war 107, 115–120; campaign model of 121, *122*, 123–129, 145, 231–236

technology 6, 34, 53, 59, 114–115, 116–117, 148–150, 206–229; as element of space power 18, 106; as limiting factor 23, 33, 48, 106; and military doctrine 63–65, 81, 87, 89, 100–105, 178; and science fiction 53

technology development 6, 41, 106, 154–155, 164, 196, 195–198; discontinuous development 111–114; Fleet Problems role in 170–179, subsidies to encourage 134, 137; systems engineering impact on 147–148

technology push 148

test of adequacy 64–65

theory 5–7

The Third Industrial Revolution 75

Tiangong (Chinese space station) 3

Titan rocket 225
titanium 19, 22, 137, 144
To Train the Fleet for War 168
Towards Distant Suns 75
trade 15–16, 24, 57, 180, 182–183, 188, 190, 208, 234, 236; mercantilist views on 129–131; trade theory 133–136; *see also* Gaumory/Baumol tarade theory
tradition 18, 64–65, 90
transformers (space power) 14, 26, *27*, 98, 164, 195–199, 232; as business plans 28–29; definition of 26; as doctrine 30–33; role in path development *42*, 43–47; role of entrepreneur 63–64; role of science fiction 53, 55, 58; as soft power 29–30
treasure 55–56, 58, 174, 188, 190
2001: A Space Odyssey 75
2010: The Year We Make Contact 75

United States 53, 79–80, 89, 92, 101, 105, 107, 143, 151, 205; historical exploration policy 82–83; merchant marine of 180–188; policy of 199; space policy of 94–97; and space power 1–5, 29, 31, 49–51, 73–74, 77, 207–210, 215, 219, 225, 233–238, 240–241; trade policy 135–136; in World War II 158, 160–161, 166, 170, 172, 177, 192–193
United States Exploring Expedition 83–84
United States Merchant Marine Academy 187, 239–240
United States Military Academy 82–83, 100
United States Naval Academy 238
United States Strategic Command (STRATCOM) 212–213
USA Space Fleet Academy *see* Space Academy

V-2 Rocket 67
vacuum *35*, 37
Valentine, Lee 241
values 64–65, 100
Van Allen, James 69–70, 114
Van Der Heijden, Kees 202–203
Veblen, Thorstein 6
Venus 213
victory 119–120, 124, 153, 232, 235
Victory ships 185
Vietnam War 234
Virgin Galactic 137, 230
Von Braun, Wernher 66–69, 71, 80
von Neumann, John 70
von Neumann probe 70
Voyager 70

war gaming 200, 235, 237; and the Fleet Problems 171, 173, 179, 195; naval gaming 161–162, 169, 176; and the Pacific War 165–166
war heresy 59–60
War Plan Orange 170, 163–174, 183, 187, 189
War Plans Division, Navy 162, 179
War Shipping Administration 186
warp speed 2, 147, 227
Washington Naval Treaty (1922) 164, 181, 187–188, 191–194, 199
wealth 2, 65; from space 15–16, 19, 24–26, 28, 34, 43, 74, 209, 229–231; and grammar of commerce 56–60; and mercantilism 129–130; and naval activity 132–135, 192; and transformers 195, 209, 229–231
weaponization of space 60, 215
weapons 63, 98, 159, 161; and doctrine development 129, 178, 200; naval weapons 166, 169, 173, 175, 181; space weapons 67, 80, 91–95, 101, 192, 206–212, 214–215, 219–220, 222, 232, 234–235; in technological warfare 121–124
West Point *see* United States Military Academy
Wheeler, George 83
Whither Space Power? 84
wild card 203
Wilson, Woodrow 184, 186
Winton, Harold 6–7
Worden, Simon "Pete" 84
World War I 162, 178–179, 197, 230; aircraft in 100–101, 145; merchant marine in 185–188; naval innovation in 169, 172–173; trench warfare in 90
World War II 7, 67, 83, 158, 162, 165, 168, 174, 184–186, 191, 197, 233; innovation in 177–179; technology development in 89–91, 117, 168; *see also* Pacific War
World War III (hypothetical) 95
Wright Brothers (Orville & Wilbur) 87, 117
Wright Flyer 90, 197
Wylie, Rear Admiral J.C, USN 6, 98

X-1 147
X-33 113
X PRIZE 136

Yeager, Chuck 147
Yoshihara, Toshi 14–15, 23, 180, 228

Zero G, Zero Tax 137
Zheng He 83